Ion Exchange Resins
Biomedical and Environmental Applications

Edited by

Inamuddin[1], Maha Khan[1], Mohammad Abu Jafar Mazumder[2,3],
Mohammad Luqman[4]

[1]Department of Applied Chemistry, Zakir Husain College of Engineering and Technology, Faculty of Engineering and Technology, Aligarh Muslim University, Aligarh-202002, India

[2]Chemistry Department, King Fahd University of Petroleum & Minerals, Dhahran 31261, Saudi Arabia

[3]Interdisciplinary Research Center for Advanced Materials, King Fahd University of Petroleum & Minerals, Dhahran 31261, Saudi Arabia

[4]Department of Chemical Engineering, College of Engineering, Taibah University, Yanbu, Saudi Arabia

Published by **Materials Research Forum LLC**
Millersville, PA 17551, USA

Published as part of the book series
Materials Research Foundations
Volume 137 (2023)
ISSN 2471-8890 (Print)
ISSN 2471-8904 (Online)

Print ISBN 978-1-64490-220-2
eBook ISBN 978-1-64490-221-9

Distributed worldwide by

Materials Research Forum LLC
105 Springdale Lane
Millersville, PA 17551
USA
https://www.mrforum.com

Manufactured in the United States of America
10 9 8 7 6 5 4 3 2 1

Table of Contents

Applications of Ion Exchange Resins in Protein Separation and Purification
J.S. Ramirez Carvajal, D. Quinteros, M. R. Romero 1

Applications of Ion Exchange Resins in Vitamins Separation and Purification
Sudipta Saha, Bidyut Saha 24

Application of Ion Exchange Resins in Protein Separation and Purification
Srijita Basumallick 39

Ion Exchange Resins for Selective Separation of Toxic Metals
Arun Kumar Pramanik, Nirmala Tamang, Abhik Chatterjee, Ajaya Bhattarai, Bidyut Saha 55

Separation and Purification of Bioactive Molecules by Ion Exchange
Rabiul Alam, Bidyut Saha 75

Ion Exchange Resins as Carriers for Sustained Drug Release
Bhavana Sampath Kumar, Junaiha Kapoor, Sandra Ravi M, Dileep Francis 93

Ion Exchange Resins for Clinical Applications
Muhammad Hassan Sarfraz, Mohsin Khurshid, Bilal Aslam, Muhammad Asif Zahoor 120

Applications of Ion Exchange Resins in Water Softening
Yu. Dzyazko 142

Keyword Index
About the Editors

Preface

In an ion exchange process, the reversible exchange of ions takes place stoichiometrically in between the stationary phase (specifically ion exchange resins) and the liquid mobile phase containing ions of interest. The principle of ion exchange thus has been utilized various in industries, environment pollution remediation technologies, food safety, water purification technologies for domestic and industrial water treatment, and biomedical devices. There are a variety of ion exchange materials including inorganic, organic, and composite ion exchangers. The ion exchange properties of these ion exchange materials are utilized for their applications in the separation and pre-concentration of metal ions, proteins, vitamins, and amino acids, kidney dialysis, water treatment, water purification, desalination, proton exchange membrane fuel cells, etc.

Ion exchange Resins: Biomedical and Environmental Applications explores in detail the application of various types of ion exchange materials in the separation and purification of protein and vitamins, selective separation of toxic metals, separation and purification of bioactive molecules, drug delivery, clinical applications, and water softening. The book will provide detailed information related to the process of ion exchange material to the environmentalist, chemical industries, water softening industries, and biomedical industries. The content of the book chapters is summarized as given below:

Chapter 1 explores the application of ion exchange techniques for protein purification. This chapter details the classification, chemical modification, and characterization of resins to improve their sensitivity and analysis of variables to carry out successful separations of proteins by IEC.

Chapter 2 discusses diverse procedures for the partition and purgation of vitamins B, C, and K through several ion exchange resins.

Chapter 3 deals with methods of chromatographic protein purification with brief mechanisms. Based on functional groups and charges ion exchange resins are categorized. The mechanism of protein purification and the effect of pH is discussed. The technical part of the application and ways to improve resolution are discussed.

Chapter 4 explores the synthesis, properties, and application of ion exchange materials. Special emphasis is given to discuss the selective separation and preconcentration of toxic heavy metal ions in industries.

Chapter 5 introduces ion-exchange chromatography (IEC) as a method for separating and purifying bioactive compounds such as polyphenols, catechin derivatives from complicated plant mixtures, proteins, minor whey protein, peptides, human C-peptide, alkaloids from Chinese medicines, plasmid DNA and carbohydrates.

Chapter 6 describes the use of ion exchange resins (IER) as a mechanism for sustained drug delivery. It explains the principles of sustained drug delivery, the mechanism of drug complexation with IERs, and the pharmacological advantages of IER- mediated drug delivery. Further, the chapter discusses the properties of some commercially employed IERs for drug delivery.

Chapter 7 discusses the clinical applications of the ion-exchange resins (IER). The role of IER in improving the properties of drug formulations is discussed. Furthermore, various drug delivery systems of IER along with the applications in targeted drug delivery and therapeutics are presented.

Chapter 8 discusses the ion exchange resins which are intended for water softening. The negative effect of hardness ions on human health and equipment is also a focus of attention. Special approaches for increasing the efficiency of water softening are also reported. These approaches involve combining ion exchange with electrodialysis or ultrasound.

Editors

Materials Research Forum LLC
https://doi.org/10.21741/9781644902219-1

Chapter 1

Applications of Ion Exchange Resins in Protein Separation and Purification

J.S. Ramirez Carvajal[1,2], D. Quinteros[3,4], M.R. Romero[1,2]*

[1] Universidad Nacional de Córdoba, Facultad de Ciencias Químicas, Departamento de Química Orgánica, Córdoba, Argentina. Edificio de Ciencias II. Haya de la Torre y Medina Allende. (5000)

[2] Consejo Nacional de Investigaciones Científicas y Técnicas (CONICET), Instituto de Investigación y Desarrollo en Ingeniería de Procesos y Química Aplicada (IPQA), Córdoba, Argentina

[3] Departamento de Ciencias Farmacéuticas, Facultad de Ciencias Químicas, Universidad Nacional de Córdoba, Ciudad Universitaria, 5000-Córdoba, Argentina

[4] Unidad de Investigación y Desarrollo en Tecnología Farmacéutica (UNITEFA), CONICET

marceloricardoromero@gmail.com*

Abstract

The advances in therapeutics based on biological macromolecules and their pharmaceutical market have stimulated the development of separation and purification techniques like ion-exchange chromatography (IEC). In this regard, the effectiveness of biological therapies depends on high-purity protein fractions. Although, there are a few differences in each protein batch due to the amplification method or the degradation effect, which cause undesirable effects on patients. In this sense, the implementation of a reliable method like IEC is fundamental, and in recent years it has been adopted as a reference technique for protein separation and purification. Therefore, this chapter details the modification of resins to improve their sensitivity and the analysis and adjustment of variables to perform a successful protein separation by IEC.

Keywords

Polystyrene Divinylbenzene, Cationic and Anionic Exchangers, Resin Characterization, Resin Functionalization, Pi of Proteins, Effect of Buffer, Effect of Support, Steps for Separation, Type of Proteins

Contents

Applications of Ion Exchange Resins in Protein Separation and Purification 1

1. **Introduction** 2

2. **Types of ion exchange resins** 5

3. **Functionalization of ion exchange resin** 7

4. **Characterization of ion exchange resin** 11

    4.1   Elemental analysis 12

    4.2   FT-IR spectra 13

    4.3   Thermogravimetric analysis 14

5. **Analysis of variables for protein IEC** 14

    5.1   Stability and pI of proteins 14

    5.2   Effect of the support on the chromatographic separation of proteins 15

    5.3   Effect of buffer and mobile phase 16

6. **Steps of protein separation by IEC** 17

7. **Types of protein purified by IEC** 18

8. **Future prospects of IEC** 19

Acknowledgments 20

References 20

1.   Introduction

Numerous pathologies arise at the post-transcriptional level whose processes facilitate the generation of mature and functional ribonucleic acid (RNA). These processes make the generation of proteins for therapeutic applications auspicious [1]. Similarly, proteins do not have a universal production, and purification processes are usually laborious and are consequently very expensive. Currently, not only is it sought to separate and obtain proteins with a high purity degree, but also it has a high performance at an industrial level, that is to say, that the purification processes are simple and affordable. At the same time, this challenge represents an opportunity that has attracted attention not only from researchers but also from the industry market, where an estimate of this year calculated turnover of the recombinant protein market will exceed 1.7 billion dollars by 2026 [2]. Within this framework, different chromatographic techniques or methods allow the purification of proteins or peptides. Thus, based on their principles, they can be classified by affinity chromatography, ion exchange, affinity with immobilized metals, and reverse phase, the latter being the most common for protein purification.

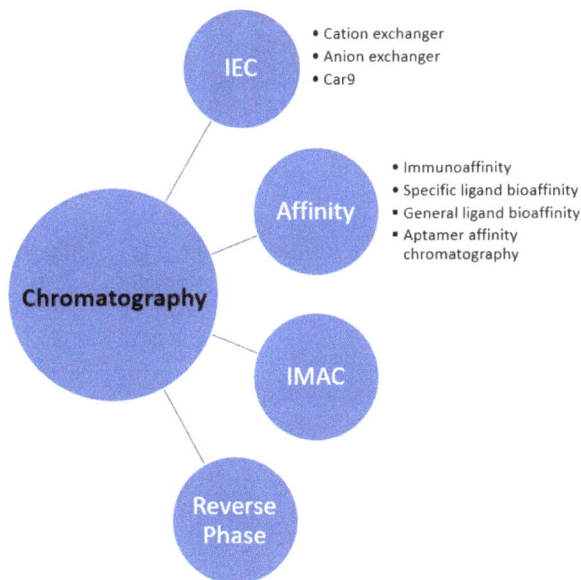

Figure 1. Chromatographic techniques applied to protein separation (from the authors).

The implementation of ion exchange resins (IER) has taken a significant impulse over the last few years (Fig. 2) since the number of scientific papers published from 2010 to 2021 is continuously growing.

The principles underlying the IER have been studied from 1950 to the present. They are mainly chemical equilibria based on ion exchange, sorption, diffusion and ion swelling, and ionization of functional groups [3]. These characteristics increase the complexity and difficulty of the study of elemental interactions. The vast number of chemical processes related to the resin phase expands the applications and possibilities for new systems.

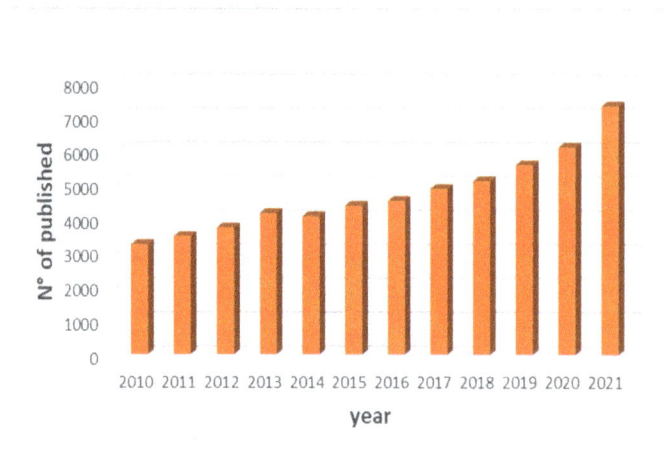

Figure 2. Progression of scientific publications related to ion exchange resins from 2010 to 2021 (source: Sciencedirect.com).

The IER has been successfully used as a catalyst for different organic compound synthesis [4-6]. Therefore, there are studies where these resins were employed to favor metrics in green chemistry processes [7].

It should be noted that the focus in IER is relevant when it is necessary to generate solutions for water treatment. The excellent plasticity of these resins' phase to adapt their chemical groups for different mechanisms of action allow the removal of a wide variety of organic and inorganic contaminants [8-12]. One of these includes the modification by adsorption processes to increase selectivity and improve their exchange properties, as found in the literature [13-15].

IER is a reference technique for protein separation and purification. In this regard, the effectiveness of biological therapies depends on high-purity protein fractions, as previously mentioned. Although, there are a few differences in each batch due to the amplification method or the degradation effect that could have undesirable effects on patients. Therefore, the application of ionic exchange for their identification, purification, and separation is essential. However, for this, a carefully variable selection is usually needed. A comprehensive approach for the analysis of different types of available substrate materials and how these variables affect the chromatographic response is detailed throughout this chapter.

Ion Exchange Resins: Biomedical and Environmental Applications   Materials Research Forum LLC
Materials Research Foundations 137 (2023) 1-23   https://doi.org/10.21741/9781644902219-1

2.   Types of ion exchange resins

IERs present unique characteristics for the remotion of heavy metals from different water sources, hydrometallurgy, wastewater treatment, biosensors, chromatography, and separations of biological macromolecules of medical interest [12]. The IER usually presents properties such as (i) temperature stability and low physical degradation, (ii) the possibility of reuse, (iii) the capacity to respond under different pH surroundings, and (iv) a simple separation method since they are insoluble in both organic and aqueous solutions [16].

The IER is often composed of a backbone based on polystyrene connected with the difunctional monomer divinylbenzene yielding a polymeric network with high chemical and structural stability, as seen in figure 3. The typical shape of manufactured resins is in particles with spherical geometry, as can be observed in the microscopic SEM image in figure 4 [17].

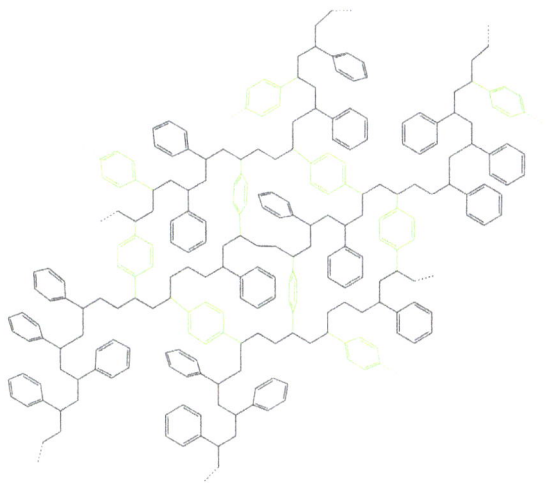

Figure 3. Scheme of chemical structure for polystyrene cross-linked with divinylbenzene. Styrene groups are colored in black and divinylbenzene in green (from the authors).

The basic resin structure can be chemically modified to expose new functional groups which allow the generation of different ion exchange capacities. For this, it is possible to establish classifications for IERs according to their particle size distribution and exchange capacity (cation or anion) depending on their specific functional groups.

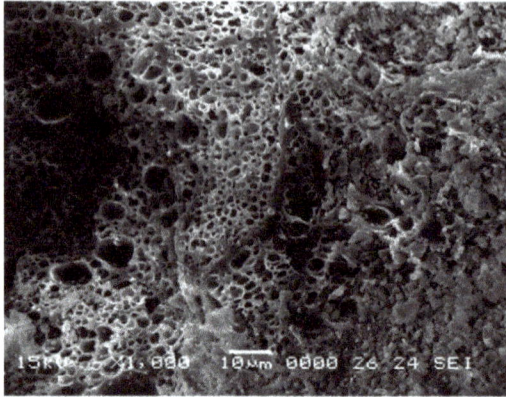

Figure 4. SEM image of QFS resin. Reprinted from [17], Copyright 2022, with permission from Elsevier.

The relationship between the type of functional groups for the exchange of certain ions and the pH dependence of the sorption capacity should be noted. Ion exchangers with strong acidity, e.g., sulfonate -SO_3H dissociate well over a wide pH range, thus reaching their highest capacity of sorption in an expanded region. On the contrary, cation exchangers containing weak acidic groups, e.g., carboxylate -COOH reach at pH> 7.0 their maximum sorption capacity (figure 5).

This characteristic should be noted as of vital importance in the correct selection of an IER for biological systems, where usually the pH must be as close to 7.0 as possible. This behavior allows the correct choice of an IER for protein purification processes, where the pH must resemble that of body fluids.

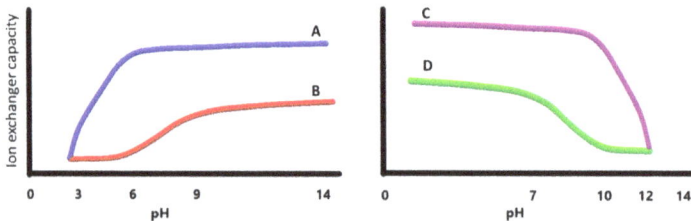

Figure 5. Sorption capacity versus pH, for cation (left) and anion (right) exchange resins. A and C are strong, B and D weak exchangers (from the authors).

The functional groups containing phosphorus like phosphonic ($-PO_3H_2$) and phosphine ($-PO_2H$), hydroxylated aromatic rings as phenol ($-C_6H_4OH$), and particularly the dicarboxylic acid amine iminodiacetate ($-N(CH_2COOH)_2$) are the most representative structures of the superficially modified resin. These groups present an acidic response and dissociate from H^+ or Na^+ ions, exchanging these species with different cations solvated in the solution. The mentioned functional groups have negative charges equivalent to the number of counter ions. These counter ions (cations) are not fixed to the negative charges and can be easily exchanged by ions belonging to the solution when these species are close enough to interact with the functional groups of the resin surface.

The third type of IERs are matrixes that contain amphoteric exchangers, which can exchange positive and negative species (cations or anions) depending on the pH of the media. These IERs are also called bipolar electrolyte exchange resins (BEE) or zwitterionic ion exchangers [18]. The morphological characteristics and chemical nature of the resin phase determines the effectiveness in the retaining process of individual ions in the sample. Furthermore, the varying degree of BEE can be controlled by selecting the composition of groups that confers the desired affinity for the resin phase. The benefit of applying this phenomenon is the efficient separation of metal ions of complex samples [19]. For further reading about the affinity series for various IERs, the work of Z. Hubicki et. al is strongly recommended [16].

3. Functionalization of ion exchange resin

The surface of IER modified with different functional groups makes the resin more or less specific to interact with different compounds of interest. For this, IERs are adapted for specific requirements by surface modification.

For metal extraction of solutions, a chloromethylated resin was functionalized with 2,2´-pyridyl imidazole to selectively adsorb and separate Ni (II) ions from other metal cations in sulfate solutions [20]. The evaluation of different factors influencing the sorption process of various metal ions on the functionalized resin allowed the selection of optimal pH and contact time conditions for the selective separation of nickel (II) ions. Under suitable conditions, the functionalized resin has a loading capacity between 56 – 79 mg of Ni (II)/g resin.

The functionalized resin was obtained by swelling the chloromethylated resin with DMF overnight at room temperature. Subsequently, 2,2´-pyridyl imidazole was added. This mixture was then refluxed for 24 h at 70 °C. The synthesized material was thoroughly washed with alcohol (preferably methanol) and purified under filtration followed by the

Soxhlet extraction process (using diethyl ether as solvent). The scheme of the reaction is shown in figure 6.

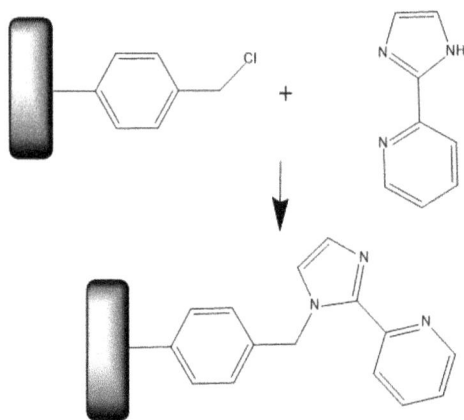

Figure 6. Schematic representation for the synthesis of a resin functionalized with 2,2'-pyridylimidazole. Based on [20], Copyright 2022, with permission from Elsevier.

There is a broad variety of ion exchange resins due to the different species that can be selected for their copolymerization or crosslinking. It is known that vinyl benzyl gel resins crosslinked with chloride/styrene/divinylbenzene, can be modified with 1,1,2–ethane tricarboxylate, 1,1–dicarboxylate–2–ethane phosphonate, and 1,1– diphosphonate–2–ethane carboxylate as shown in Figure 7 [13]. Note that the objective here is improving the acidity of the exchange agent, allowing a better interaction with the species of interest, and enhancing the selectivity.

The study of these resins reveals that modification with 1,1-dicarboxylate-2-ethanol phosphonate is an efficient way to prepare ion exchange resins. This resin shows high synthesis yields and has a strong affinity for divalent metal cations. An example of this feature is the efficient removal of Pb (II) from acidic solutions. Resin 2 was obtained with a high yield, however, the presence of only carboxyl groups on its surface makes it less efficient than resin 1. Resin 3 was obtained with a low yield and its exchange capacity was lower than resin 2.

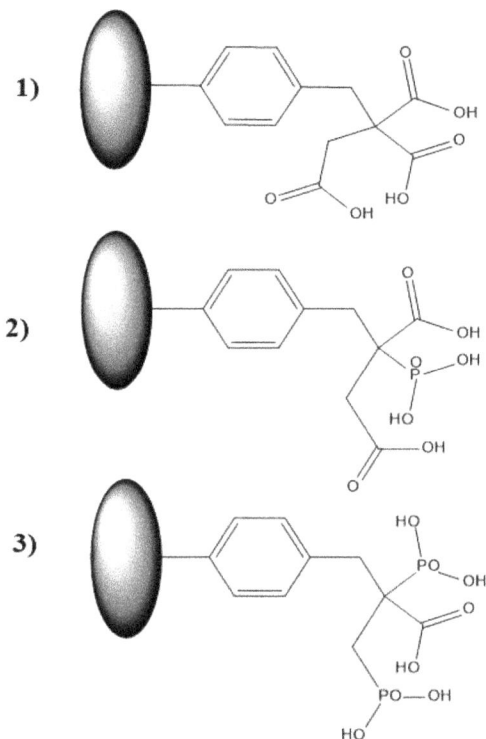

Figure 7. Structure of modified resins. Based on [13], Copyright 2022, with permission from Elsevier.

IER has been implemented in the study of the adsorption of compounds of biological and environmental importance due to their toxicity and hazardousness. The chemical structures of IER functionalized with four linear polyamines for the adsorption of uranium (in uranyl form) are shown in Figure 8. The modification of these materials improves the capacity to adsorb this contaminant in comparison with commercial resin. The studies showed that the resin functionalized with the longest polyamine chain has a maximum uranyl loading capacity (269.50 mg/g) [21].

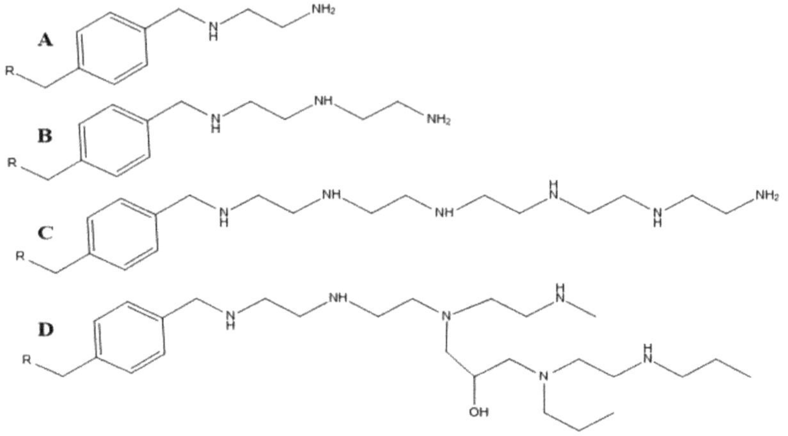

Figure 8. Typical functional groups of resins: (A)Ethylenediamine (EDA), (B) Diethylenetriamine (DETA), (C) Pentaethylenehexamine (PEHA), (D) Commercial Purolite S985. R represents the basic IER structure. Based on [21], Copyright 2022, with permission from Elsevier.

The synthesis procedures for obtaining functionalized resins involve a relatively simple method. Previously, it was required to swell the base resin for 24 h with 250 mL of 1,4-dioxane. Subsequently, 16 g of EDA and 230 mL of additional 1,4-dioxane were added to be refluxed for 48 h in a nitrogen atmosphere. Then, the product was washed with ethanol, a mixture of triethylamine: dichloromethane (1/10), and deionized water, until neutral pH is yielded. A similar procedure can be followed to functionalize with the other polyamines [21].

Resins containing styrene cross-linked with divinylbenzene as a support material were functionalized with diverse amino groups as shown in Table 1 [22]. These materials were evaluated by an O_2 absorption study.

Table 1. IER of Polystyrene-divinylbenzene (PSDVB) modified with amines. Reprinted from [22], Copyright 2022, with permission from Elsevier.

IERs	Amine group	Particle size (mm)	Support materials
VPOC 1065	Primary amine	0.315 – 1.25	PSDVB
A109	Primary amine	0.425 – 1	PSDVB
D201	Quaternary amine	0.315 – 1.25	PSDVB
D202	Quaternary amine	0.315 – 1.25	PSDVB
A830	Complex amine	0.3 – 1.2	PSDVB

Exchange capacity, surface area, and pore volume variations are typical properties dependent on and adjusted by surface modification of IER, as shown in Table 2. The study and understanding of these properties are crucial for governing their response against a specific analyte.

Particularly, VPOC 1065 shows the ability to approximately adsorb 1.05 mmol CO_2/g. Additionally, after 275 adsorption-desorption cycles, the material showed a 4.77% decrease in its adsorption capacity. This demonstrates its high stability and high reusability.

Table 2. Sorption properties of IERs after drying at a constant weight under an N_2 atmosphere. Reprinted from [22], Copyright 2022, with permission from Elsevier.

Ion exchange resins	Ion exchange capacity	BET surface area (m^2/g)	Pore volume (cm^3/g)
VPOC 1065	2.2 (eq/L)	24.5811	0.202653
A109	1 (eq/L)	25.5314	0.033331
D201	3.7 (mmol/g)	9.0443	0.015320
D202	3.6 (mmol/g)	7.1163	0.013251
A830	2.75 (eq/L)	4.4135	0.009138

4.   Characterization of ion exchange resin

There are a variety of analytical techniques available to characterize the physical and chemical properties of IERs. These methods allow analyzing the composition of a given resin and the surface modification process, which allows projecting its usefulness in a given ion exchange process. In addition to the typical morphological characterization methods such as SEM, TEM and optical or confocal microscopy there are physicochemical techniques that are detailed in this section. X-ray spectroscopy, Infrared spectroscopy, elemental analysis, and thermogravimetric methods can also be implemented to predict the coordination of an analyte on the resin surface, among other properties, allowing the development of better exchange models between the analytes and IER.

4.1 Elemental analysis

Elemental analysis (EA) of an ion exchange resin can be exploited to understand how the modifying agent binds to the surface of the material [21,23]. In this work, IER was functionalized with three linear polyamines and the percentages of nitrogen and chlorine in the samples were determined by EA, as shown in Table 3 [21].

The chlorine atoms are associated with the possible bond points on the resin surface. The results obtained allowed inferring that the polyamine molecules are arranged as cross-linked structures among themselves, and their bonding to benzyl groups.

Table 3. Elemental analysis (N and Cl) in the Merrifield resin (MR), Ps-EDA, Ps-DETA, and Ps-PEHA. Reprinted from [21], Copyright 2022, with permission from Elsevier.

	N (%)			Cl (%)		
	T	M	Y (%)	T	M	Y (%)
Commercial	0.00	0.0	-	19.5	22.7	-
Ps-EDA	17.90	9.82	54.9	0	2.18	90.38
Ps-DETA	26.85	11.54	43.0	0	2.41	89.36
Ps-PEHA	53.69	13.13	24.5	0	5.00	77.92

T= theoretical, M= measured, and Y= yield

The lower percentage of nitrogen in Ps-PEHA compared to Ps-EDA and Ps-DETA indicates that the size of the modifying molecule influences the functionalization of the resin surface. The elemental analysis provides insight regarding the reduction of available sites for functionalization based on the restriction of the movement of species through the resin pores. Thus, elemental analysis of the IER is a valuable tool to analyze the best conditions for the functionalization of an IER [21].

4.2 FT-IR spectra

The FT-IR spectra are typically implemented for verification of the modification and reusability of the IER.

The FT-IR spectra reveal the specific vibrational signals of resins. Figure 9 shows commercial and three functionalized resins with different polyamines [21]. In this work, this technique implementation aims to confirm the presence of the modifying molecules on the IER surface.

Since all materials have a similar skeleton, all spectra present common signals. The differences found in the spectra allow us to infer that the functionalization was successful and that it was through a nucleophilic substitution, in which the amine replaces the chlorine. This is evidenced by the disappearance of the C-Cl stretching signal at 673 cm^{-1} in the functionalized resins. Thus, resin characterization by FT-IR spectra allows a qualitative evaluation of functionalization processes.

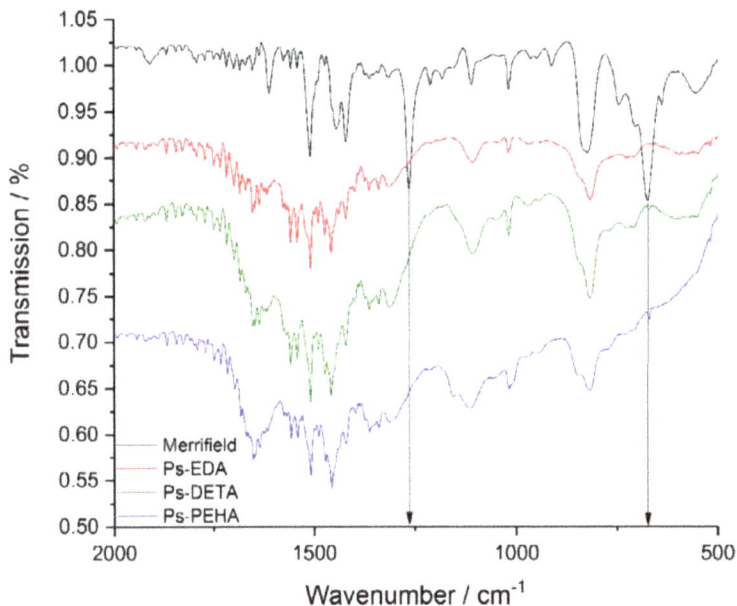

Figure 9. Region of interest selected of the FT-IR spectra of the MR, Ps-EDA, Ps-DETA, and Ps- PEHA. Arrows indicate the C-Cl and CH$_2$Cl stretches (673 cm^{-1} and 1265 cm^{-1}). Reprinted from [21], Copyright 2022, with permission from Elsevier.

FT-IR spectra can be exploited to determine the reusability of IERs. FTIR spectra of a resin sensitive to CO_2 before and after the absorption process can be used to characterize the changes in the material [22].

This analysis allows researchers to follow the structural behavior of the resin after an absorption-desorption cycle of CO_2. This method is important because it is possible to study the sample after a large number of absorption-desorption cycles without generating significant changes in the structure of the material.

4.3   Thermogravimetric analysis

Thermogravimetric analysis (TGA) allows the characterization of mass loss of IER during the exposure of the material to a temperature ramp under an air or nitrogen atmosphere. An exhaustive analysis of the degradation ranges allows researchers to consider the possible species formed on the surface when functionalizing an exchange resin [21].

Another way to exploit TGA is the evaluation of thermal degradation characteristics, as well as calculating kinetic parameters such as activation energy and reaction order [23].

5.   Analysis of variables for protein IEC

5.1   Stability and pI of proteins

The pI of proteins is the point where the sum of positive and negative charges is equivalent, and in this zwitterionic state, the net charge is zero. The pI can be calculated by knowing the amino acid sequence of the protein and some online calculators automatically can perform this task [24]. However, it should be noted that these estimated values are approximate since the microenvironment surrounding each acid-base group influences the actual pI value.

As a consequence, when the protein is at pI, it has a low affinity for the exchange resin and is better eluted under these conditions. Oppositely, at pH above pI, it binds more efficiently to anionic exchangers while below it to cationic ones. The stability of a protein is related to pI, for this reason, it cannot be exposed to pH very different from this value without producing irreversible changes in the conformation or aggregation that induce precipitation. In practice, the pH of the medium should not be more than 2 units away from the pI of the target proteins in the sample.

In IEC, several parameters are dependent on temperatures, such as the retention factor, viscosity, and pH of the buffer. However, when the sample is a protein, the tertiary and quaternary structures are the most susceptible factors. For this reason, before carrying out separation by ion exchange, the stability of the protein solution should be tested, exposing small fractions of it to temperatures from 4 to 40°C. In a complementary assay, stability should be verified against pH (between 2 and 10), salt concentrations (up to 1M with NaCl), and the presence of remnant proteases (exposing the sample to room temperature overnight and evaluating the concentration of the supernatant by UV at 260-280nm).

5.2 Effect of the support on the chromatographic separation of proteins

As mentioned in the previous section, there are strong and weak exchangers, both cationic and anionic. The strong ones remain charged over a wide pH range, and the weak ones vary their charge over a short pH range. For this reason, to start with a sample of unknown proteins, it is preferable to work with a strong exchanger, especially in extreme pH conditions. Although, if the selectivity is low, it may be necessary to work with the weak type. Within the strong exchangers, cationic ones adjust better to the main systems, but when the PI of the proteins is less than 7, the use of an anion exchanger is strongly suggested.

There are different matrices based on silica. However, in recent years, they have been widely surpassed by polymers based on modified divinylbenzene since they withstand more flexibility in working conditions, as was detailed in the preceding paragraphs.

The relation of chromatographic column parameters was first modeled by Jan Jozef van Deemter. The van Deemter equation relates the height equivalent to a theoretical plate (HEPT) with the kinetic and flow parameters as shown in equation 1 [25]. These parameters allow to determine and select the appropriate porosity of support of interest:

$$HEPT \; = A + \left(\frac{B}{u}\right) + C \, u \tag{1}$$

Where HEPT is related to the resolving power (m), A is the parameter of Eddy diffusion (m), B is the longitudinal diffusion coefficient of the particles (m^2s^{-1}), u is the mean velocity of the mobile phase (ms^{-1}) and C is the coefficient of resistance to mass transport of the analyte (s). The term C related to mass transfer in liquid chromatography can be interpreted as follows (equation 2):

$$C \; \propto \frac{d^2\gamma}{D} u \tag{2}$$

Where d is the particle diameter of the support, γ is the tortuosity and D is the diffusivity of the mobile phase. In this case, protein samples, molecules with high molecular weight and low diffusivity, are strongly affected by γ, causing band broadening. For this reason, smaller particle sizes, high crosslinking, and low porosity supports are preferred at the expense of lower retention and capacity.

5.3   Effect of buffer and mobile phase

The adequate protein separation depends on the precise adjustment of buffer composition. In general, a buffer with pH between the pI of the macromolecule and the pKa of the acid or base present in the cation or anion exchanger is preferable. In addition, it is not more than one pH unit away from the pI to avoid its denaturation. As shown in figure 10, when there is a strong cation exchanger (acidic pKa), the pH of the buffer must be lower than the pI of the protein, which has an excess of positive charges, favoring its interaction with the support. On the contrary, when there is an anion exchanger whose pKa is high, the pI should be lower than the pH of the chosen buffer. In this case, the protein net charge is negative, achieving a better interaction with the support.

The buffer concentration is usually between 10 and 50 mM. Although, when the buffer concentrations are low, the proteins could interact very strongly with the support, preventing their elution. In addition, an elevated quantity of species in the solution could cause denaturation. The molecules used are common ones such as phosphate, citrate, (N-morpholino) ethane sulfonic acid (MES), and acetate, among others [26]. This type of buffer, called typically starting buffer, is employed to favor the binding of the protein to the support.

After the interaction occurs and the non-adsorbed and contaminating molecules are washed away, the elution step can proceed. The composition and nature of the elution buffer will depend on the method used to reduce the interaction of the protein with the support. Thus, if the selected methodology is by salt gradient, the composition of the buffer will not change. Conversely, it will incorporate an increasing amount of salt (generally sodium or potassium chloride between 0.2 and 0.5 M) to reduce the interactions of the protein with the support and thus achieve its elution. On the other hand, if the elution is by pH, the elution buffer must be adjusted to a pH close to the pI of the protein, being higher than pI in cation exchangers or lower in anion exchangers. Following the reverse analysis of what was discussed above, in this stage, the elution of the molecule of interest is sought.

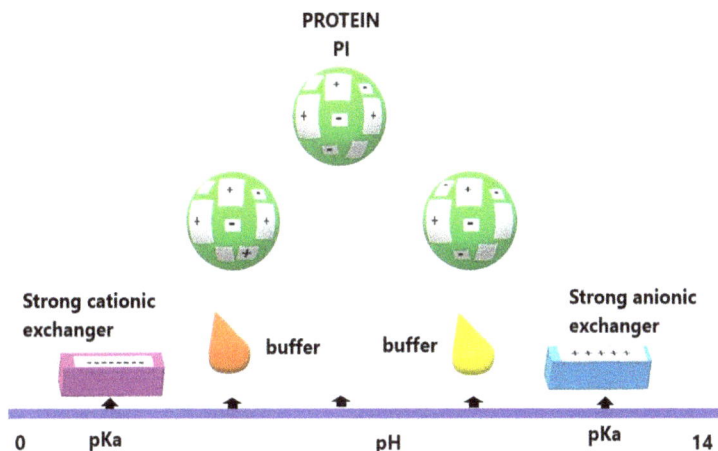

Figure 10. Effect of starting buffer pH in protein charge balance with strong cationic and anionic exchangers (from the authors).

6.   Steps of protein separation by IEC

Figure 11 shows a typical ion exchange protein separation using the ionic gradient elution method. For this, the first step consists of conditioning, also the solutions and the column must be at the same temperature to avoid the formation of bubbles within the system. The amount of sample to pass through the column should be 20 to 40% of the manufacturer's specified capacity. Subsequently, 5 to 10 times the column volume (CV) of the starting buffer is passed through it until the pH, conductivity, and baseline in the detector are stable. Subsequently, the pH and ionic strength of the problem sample are adjusted as close as possible to the starting buffer. The sample is passed using 5 to 10 CV until the signals from the unbound sample components disappear and the initial baseline is stable on the detector. Then, depending on the selected methodology, elution can be either by gradient or steps selecting the adjustment of pH or by ionic strength. Typically, 15 to 20 CVs of an elution buffer composed of a starting buffer containing a salt concentration gradient up to a maximum value of 500 mM are passed during the run. Starting buffer steps can be made with stepping additions of constant salt at concentrations chosen for elution of the protein of interest. Column regeneration is performed with 5 CV of 1 M NaCl for elution of the remaining bonded components. Finally, the support can then be rebalanced with 5-10 CV of starting buffer until observing the baseline is reached again. The flow rate depends on

17

the type of column but in normal conditions is 1 mL/min, varying from 0.1 to 2 mL/min according to their different characteristics. It also depends on the chromatography stage, since the column preparation processes, as well as the regeneration processes, are carried out at higher flows, while the sample loading and elution processes are performed at a slower speed. More details on the exposed topic can be found in the Cytiva IEC Handbook [27].

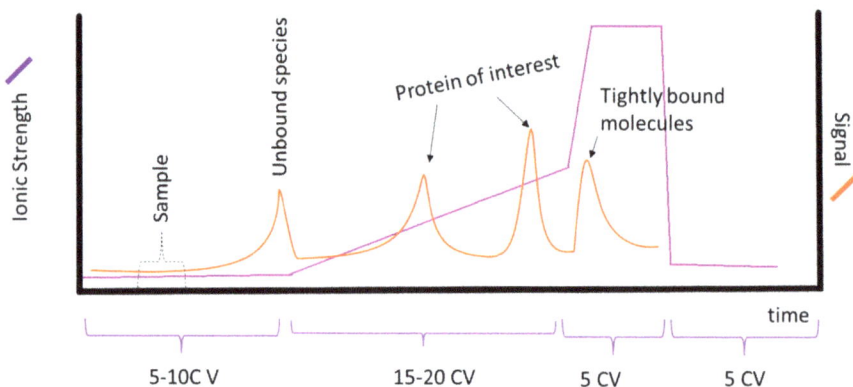

Figure 11. Typical ion exchange protein separation using ionic gradient elution method (from the authors).

7. Types of protein purified by IEC

The use of specific tags is the preferable strategy for large-scale protein production and purification as it allows efficient entrapment in chromatography columns and can affect protein stability and functionality [28]. The correct use of these labels can help reduce the costs related to preparing chromatography columns since from the same chromatography, it is possible to separate and purify one or more proteins, applying to large-scale production. The use of protein and peptide labels has revolutionized the market due to their low cost as well as for the purification of high-impact proteins in the pharmaceutical industry, mainly for the mass production of antiviral vaccines such as Provenge® and Flublok® (systems of baculovirus expression vectors) or the recombinant protein subunits for vaccines containing antiviral antigens without being pathogenic and presenting minimal risk.

Every protein or peptide can have a variety of residues, by the function of pH. The separation of proteins by ion exchange chromatographic columns is an excellent option since being formed by solid resins they have both positive ion-exchange groups such as tertiary and quaternary ammonium cations (anion exchangers) and negative such as sulfonic and carboxylic acids (cation exchangers).

(i) Cation exchangers. Studies conducted by Hedhammar et al. [29] showed that the realization of a structure formed by three helices of 58 amino acids derived from Staphylococcus protein A allowed fusing with different target proteins obtaining a specific purification of the desired protein.

(ii) Anion exchangers. On the cell surface, there is a great number of proteins. Their charge is predominantly negative, and as a consequence, positive ion exchange IEC is considered a suitable tool for protein separation and purification in many biotechnological and biomedical studies. A typical complementary method of purification includes IEC and affinity with Z-basic domain derived from Staphylococcus protein A [30], showing that the combination is an excellent preservation of biomolecule activity.

In addition, we would like to mention how, through an ion exchange film technique, it has been possible to eliminate endotoxins during the manufacturing process. It is essential to guarantee drug quality and patient safety [31]. The procedure was achieved using anion exchange chromatography, while partial removal of endotoxins was obtained with cation exchange chromatography.

Coyle et al. reported an IER cationic based on silica gel-filled (Whatman DE52) combined with the peptide Tag Car9 (TC) to obtain an efficient protein purification. This IER could be easily disrupted using concentrated amino acid solutions (e.g., L-arginine or L-lysine), and the short purification period with adequate proteolytic susceptibility results in biomolecules with a C-terminus, and an even significantly reduced purification cost [32,33].

8.   Future prospects of IEC

It is clear that treatment methods with complex molecules such as antibodies are just beginning and will continue to grow for many years. This driving force will lead to the development of better and more efficient separation techniques such as those based on IEC. With the development of powerful computational methods, they will assist with increasing relevance in the design of support materials and conditions tailored for each separation molecule, minimizing the need for tests and even optimizations based on the design of experiments.

Acknowledgments

J.S. Ramirez Carvajal acknowledges CONICET for the postdoctoral fellow received. Financial support from FCQ-UNC and IPQA – CONICET (Argentina) is gratefully acknowledged. We also like to thank Prof. Inamuddin for his invitation.

References

[1] B. Leader, Q. Baca, D. Golan, Protein therapeutics: a summary and pharmacological classification. Nat. Rev. Drug Discov. 7 (2008) 21-39. https://doi.org/10.1038/nrd2399

[2] Recombinant Protein Market, Report Code: BT 7839. https://www.marketsandmarkets.com/Market-Reports/recombinant-proteins-market-70095015, 2021 (accessed 1 April 2022).

[3] V.F. Selemenev, A.A. Zagorodni, Infrared spectroscopy of ion-exchange resins. Determination of amino acids ionic form in the resin phase, React. Funct. Polym. 39 (1999) 53-62. https://doi.org/10.1016/S1381-5148(97)00173-9

[4] R.C.F. Zeferino, V.A.A. Piaia, V.T. Orso, V.M. Pinheiro, M. Zanetti, G.L. Colpani, N. Padoin, C. Soares, M.A. Fiori, H.G. Riella, Synthesis of geranyl acetate by esterification of geraniol with acetic anhydride through heterogeneous catalysis using ion exchange resin, Chem. Eng. Res. Des. 168 (2021) 156-168. https://doi.org/10.1016/j.cherd.2021.01.031

[5] R.C.F. Zeferino, V.A.A. Piaia, V.T. Orso, V.M. Pinheiro, M. Zanetti, G.L. Colpani, N. Padoin, C. Soares, M.A. Fiori, H.G. Riella, Neryl acetate synthesis from nerol esterification with acetic anhydride by heterogeneous catalysis using ion exchange resin, J. Ind. Eng. Chem. 105 (2022) 121-131. https://doi.org/10.1016/j.jiec.2021.09.015

[6] N. Xia, W. Wan, S. Zhu, H. Wang, K. Ally, Synthesis and characterization of a novel soluble neohesperidin-copper (II) complex using Ion-exchange resin column, Polyhedron. 188 (2020) 114694. https://doi.org/10.1016/j.poly.2020.114694

[7] J.H. Badia, E. Ramírez, R. Soto, R. Bringué, J. Tejero, F. Cunill, Optimization and green metrics analysis of the liquid-phase synthesis of sec-butyl levulinate by esterification of levulinic acid with 1-butene over ion-exchange resins, Fuel Process. Technol. 220 (2021) 106893. https://doi.org/10.1016/j.fuproc.2021.106893

[8] Y. Öztürk, Z. Ekmekçi, Removal of sulfate ions from process water by ion exchange resins, Miner. Eng. 159 (2020) 106613. https://doi.org/10.1016/j.mineng.2020.106613

[9] S. Cheng, J. Qian, X. Zhang, Z. Lu, B. Pan, Commercial Gel-Type Ion Exchange Resin Enables Large-Scale Production of Ultrasmall Nanoparticles for Highly Efficient Water Decontamination, Engineering. (2021). https://doi.org/10.1016/j.eng.2021.09.010

[10] E. Çermikli, F. Şen, E. Altıok, J. Wolska, P. Cyganowski, N. Kabay, M. Bryjak, M. Arda, M. Yüksel, Performances of novel chelating ion exchange resins for boron and arsenic removal from saline geothermal water using adsorption-membrane filtration hybrid process, Desalination. 491 (2020) 114504. https://doi.org/10.1016/j.desal.2020.114504

[11] R. de Abreu Domingos, F.V. da Fonseca, Evaluation of adsorbent and ion exchange resins for removal of organic matter from petroleum refinery wastewaters aiming to increase water reuse, J. Environ. Manage. 214 (2018) 362-369. https://doi.org/10.1016/j.jenvman.2018.03.022

[12] Y.A.R. Lebron, V.R. Moreira, M.C.S. Amaral, Metallic ions recovery from membrane separation processes concentrate: A special look into ion exchange resins, Chem. Eng. J. 425 (2021) 131812. https://doi.org/10.1016/j.cej.2021.131812

[13] W.A. Trochimczuk, New ion-exchange/coordination resins with carboxylate and phosphate functional groups, Eur. Polym. J. 35 (1999) 1457-1464. https://doi.org/10.1016/S0014-3057(98)00220-1

[14] H. Liu, S. Ning, S. Zhang, X. Wang, L. Chen, T. Fujita, Y. Wei, Preparation of a mesoporous ion-exchange resin for efficient separation of palladium from simulated electroplating wastewater, J. Environ. Chem. Eng. 10 (2022) 106966. https://doi.org/10.1016/j.jece.2021.106966

[15] E. Lafond, C. Cau Dit Coumes, S. Gauffinet, D. Chartier, P. Le Bescop, L. Stefan, A. Nonat, Investigation of the swelling behavior of cationic exchange resins saturated with Na+ ions in a C3S paste, Cem. Concr. Res. 69 (2015) 61-71. https://doi.org/10.1016/j.cemconres.2014.12.009

[16] Z. Hubicki, D. Koodynsk, Selective Removal of Heavy Metal Ions from Waters and WasteWaters Using Ion Exchange Methods, in: Ayben Kilislioğlu (Ed.), Ion Exchange Technology, Intech Open, London, 2012, pp. 193-240. https://doi.org/10.5772/51040

[17] R.S. Azarudeen, M.A.R. Ahamed, A.R. Burkanudeen, Chelating terpolymer resin: Synthesis, characterization and its ion-exchange properties, DES. 268 (2011) 90-96. https://doi.org/10.1016/j.desal.2010.10.002

[18] O.N. Kononova, A.G. Kholmogorov, S. V. Kachin, O. V. Mytykh, Y.S. Kononov, O.P. Kalyakina, G.L. Pashkov, Ion exchange recovery of nickel from manganese nitrate solutions, Hydrometallurgy. 54 (2000) 107-115. https://doi.org/10.1016/S0304-386X(99)00052-3

[19] J.S. Fritz, Factors affecting selectivity in ion chromatography, J. Chromatogr. A. 1085 (2005) 8-17. https://doi.org/10.1016/j.chroma.2004.12.087

[20] A.I. Okewole, E. Antunes, T. Nyokong, Z.R. Tshentu, The development of novel nickel selective amine extractants: 2,20′-pyridyl imidazole functionalized chelating resin, Miner. Eng. 54 (2013) 88-93. https://doi.org/10.1016/j.mineng.2013.04.019

[21] J.T.M. Amphlett, M.D. Ogden, R.I. Foster, N. Syna, K. Soldenhoff, C.A. Sharrad, Polyamine functionalized ion exchange resins: Synthesis, characterization and uranyl uptake, Chem. Eng. J. 334 (2018) 1361-1370. https://doi.org/10.1016/j.cej.2017.11.040

[22] M. Parvazinia, S. Garcia, M. Maroto-Valer, CO2 capture by ion exchange resins as amine functionalized adsorbents, Chem. Eng. J. 331 (2018) 335-342. https://doi.org/10.1016/j.cej.2017.08.087

[23] M.N. Yasin, R.K. Brooke, S. Rudd, A. Chan, W.T. Chen, G.I.N. Waterhouse, D. Evans, I.D. Rupenthal, D. Svirskis, 3-Dimensionally ordered macroporous PEDOT ion-exchange resins prepared by vapor phase polymerization for triggered drug delivery: Fabrication and characterization, Electrochim. Acta. 269 (2018) 560-570. https://doi.org/10.1016/j.electacta.2018.02.162

[24] Isoelectric point calculator, http://www.calistry.org/calculate/isoelectric-point-calculator (Accessed 5 May 2022).

[25] J.J. Van Deemter, F.J. Zuiderweg, A. Klinkenberg, Longitudinal diffusion and resistance to mass transfer as causes of non-ideality in chromatography, Chem. Eng. Sc. 5 (1956) 271-289. https://doi.org/10.1016/0009-2509(56)80003-1

[26] S. Fekete, A. Beck, J.L. Veuthey, D. Guillarme, Ion-exchange chromatography for the characterization of biopharmaceuticals, J. Pharm. Biomed. Anal. 113 (2015) 43-55. https://doi.org/10.1016/j.jpba.2015.02.037

[27] IEC Handbook, https://www.cytivalifesciences.com/es/ar/support/handbooks (Accessed 20 May 2022).

[28] D. Guan, Z. Chen, Challenges and recent advances in affinity purification of tag-free proteins. Biotechnol. Lett. 36 (2014) 1391-1406. https://doi.org/10.1007/s10529-014-1509-2

[29] M. Hedhammar, T. Graslund, M. Uhlen, S. Hober, Negatively charged purification tags for selective anion-exchange recovery. Protein Eng. Des. Sel. 17 (2004) 779-786.

[30] Hedhammar, M., Hober, S., 2007. Zbasic-A novel purification tag for efficient protein recovery. J. Chromatogr. A 1161 (1-2), 22-28. https://doi.org/10.1016/j.chroma.2007.05.091

[31] M. Wespel, M. Geiss, M. Nägele, S. Combé, J. Reich, J. Studts, J. Stolzenberger, The impact of endotoxin masking on the removal of endotoxin during manufacturing of a biopharmaceutical drug, Journal of Chromatography A. 1671 (2022) 462995. https://doi.org/10.1016/j.chroma.2022.462995

[32] B.L. Coyle, F. Baneyx, A cleavable silica-binding affinity tag for rapid and inexpensive protein purification, Biotechnol. Bioeng. 111 (2014) 2019-2026. https://doi.org/10.1002/bit.25257

[33] M. Xu, M.J. Bailey, J. Look, F. Baneyx, Affinity purification of Car9-tagged proteins on silica-derivatized spin columns and 96-well plates, Protein Expr. Purif. 170 (2020) 105608. https://doi.org/10.1016/j.pep.2020.105608

Materials Research Forum LLC
https://doi.org/10.21741/9781644902219-2

Chapter 2

Applications of Ion Exchange Resins in Vitamins Separation and Purification

Sudipta Saha[1], Bidyut Saha[2]*

[1]Assistant Professor, Department of Chemistry, Triveni devi Bhalotia College, Raniganj, Paschim Bardhaman, West Bengal, India-713347

[2]Professor, Department of Chemistry, The University of Burdwan, Burdwan, West-Bengal, India-713104

Abstract

Vitamins are vital compounds for the improved development and growth of the human body system. Together chemically and analytically, vitamins are divergent assembly of composites as they comprise of biomolecules. The source of vitamins is the necessary dietary food stuff. Small amounts of vitamins are needed to sustain a healthy life. Vitamins are generally classified into fat-soluble and water-soluble vitamins. The major portions of vitamins are vitamin A, vitamin B, vitamin C, vitamin E and vitamin K. This chapter discussed diverse procedures for the partition and purgation of vitamins B, C and K through several ion exchange resins.

Keywords

Vitamins, Water Soluble Vitamins, Fat Soluble Vitamins, Ion Exchange Resins, Separation, Purification

Contents

Applications of Ion Exchange Resins in Vitamins Separation
and Purification...24

1.   Introduction...25

2.   Importance of vitamins 26

3.   Categorisation of vitamins 26

        3.1   Water soluble vitamins 26

3.2   Fat soluble vitamins 27

4.   Origin of vitamins...**27**

5.   Isolation and purgation of vitamin **29**

6.   Ion-exchange chromatography..**29**

7.   Ion exchange chromatographic isolation and purgation of vitamin K1 ....**30**

8.   Ion exchange chromatographic isolation and purgation of vitamin C...**30**

9.   Ion exchange chromatographic isolation and purgation of vitamin B1, vitamin B2 and vitamin B6..**31**

Conclusion...**32**

References **32**

1.   Introduction

Enzymatic and chemical reactions of the human body are not possible without vitamins. The human body cannot synthesise vitamins naturally. Vitamins are taken from balanced foodstuffs to maintain the proper function and metabolism of our physiological system on an everyday basis [1-6]. The inadequate supply of vitamins leads to various diseases such as metabolic disorders, cardiac problems, diabetes and tumours also [1,6-12]. Vitamins are categorised into water soluble and fat-soluble vitamins. Vitamin B and C are water-soluble vitamins. On the other hand, Vitamins A, D, E, K are fat soluble vitamins [1,13-17]. Animal food stuffs such as egg, milk, meat are major sources of vitamins. Numerous procedures including biological technique, chemical technique, and physical techniques are employed to isolate and refine vitamins [1,18-21].

The separation and purification of fat-soluble and water-soluble vitamins are carried out by various chromatographic techniques such as HPLC, ion exchange chromatography and also by Capillary Electrophoresis [1,22-26]. The ion exchange resin has a vital part in the many environmental applications such as wastewater treatment, catalysis, hydrometallurgy, ecological remediation, water softening, biomolecular partitions. The ion exchange chromatography is generally used magnificently for vitamin separation and purification. A variety of synthetically prepared and farther resins are engaged to serve this purpose [1,27-28]. The ion exchange chromatography isolates vitamins mainly by cation and anion exchange resins. Demineralization, conversion, purification and concentration are the major steps of the ion exchange chromatography [1,28-29].

2.   Importance of vitamins

Vitamins are inevitable biomolecules for our healthy human body system on a daily basis. Improper supply of vitamins is responsible for serious physiological disorders. The human body system is unable to produce vitamins inherently. Vitamins are taken from healthy and balanced diets. Many serious diseases like diabetes, heart diseases, tumour, metabolic illness are the result of improper intake of vitamins. In the current era, due to evolution and development, many creatures have lost their power to produce vitamins. In this connection, Homo sapiens cannot arrange the vitamins which are extremely required for the existence and development of the human body system. All the different types of vitamin B i.e., vitamin B1, vitamin B2, vitamin B3, vitamin B6 and vitamin B12 repair cells and support ingestion. These said vitamins improve the metabolic and immunity power of the human body system. Vitamin C also lifts the immunity power and aids in the development of individuals. Moreover, it lowers the hazard of elevated blood pressure, anaemia, cholesterol and numerous further illnesses. Vitamin K exhibits a crucial role in the thrombosis of the human body system. It also sustains the robust bones [1,17].

3.   Categorisation of vitamins

Total 13 types of vitamins are classified into water soluble and fat-soluble vitamins. This classification is based on their solubility power into fats and water. Vitamin B complex and Vitamin C fall into the water-soluble vitamins. On the other hand, the vitamins A, D, E, K are the fat-soluble vitamins [30].

3.1   Water soluble vitamins

The aqueous system easily emulsifies the water soluble vitamins. The B group vitamins such as B1, B2, B3, B6 and B12 and vitamin C are the water soluble vitamins. The Vitamin C, riboflavin, and thiamine are extremely sensitive to warmth and in the alkaline condition, these vitamins are highly unstable and possess a complex structure. This causes the separation and purification of water-soluble vitamins hard to accomplish. The discharge of these types of vitamins from anatomy is performed during urination. Proper storage of water-soluble vitamins is tough as they are degraded and washed off effortlessly. These vitamins act as coenzymes in various chemical reactions. They also improve the energy metabolism [31-33].

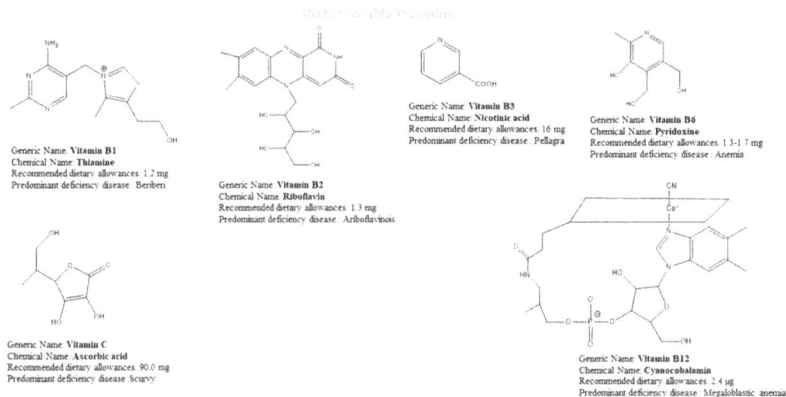

Fig.1.0. Names and structures of the Water Soluble vitamins, their chief scarcity illnesses and the suggested dietary doses

3.2   Fat soluble vitamins

The Vitamins A, D, E, K are the fat-soluble vitamins. These vitamins are water insoluble but immediately dissolve into fat. Liver stores the fat-soluble vitamins and excess intake of these vitamins leads to serious health problems because the excretion rate of these vitamins is lower. The transmission and metabolism of these vitamins needs fatty materials with high lipid content. These vitamins are easily adsorbed through dietary fats. The proper circulation of bile and generation of micelle increases the adsorption power of fat-soluble vitamins. The transport of fat-soluble vitamins starting from the lymphatic tissue towards the hepatic organ is assisted by chylomicrons. The lack of these vitamins leads to various physiological disorders like tumour, diabetic mellitus type-II and immune systems [34-37].

4.   Origin of vitamins

The plant kingdom is a major supplier of vitamins. The B group vitamins are found in vegetables, fruits (spinach, potatoes, broccoli, asparagus, bananas, dates) seeds, figs, pulses, nuts. They are also found in dairy products. Alternatively, citrus fruits supply a major part attributed to the ascorbic acid i.e., vitamin C. Besides this, vitamin C is also found in vegetables (potatoes, green pepper, spinach, broccoli), lemons, strawberries, peas, pears, lime, in animal-based products like pork, chicken and also in seafood. Vitamin K is also seen in some vegetable-based sources like soya bean, cucumbers, asparagus, Apium graveolens dulce, coriander, head lettuce, sauerkraut, kale, brussels sprouts, whole grain, peas. The animal-based sources of vitamin K are liver, flesh and albumen (Table-1) [1,34-37].

Fat Soluble Vitamins

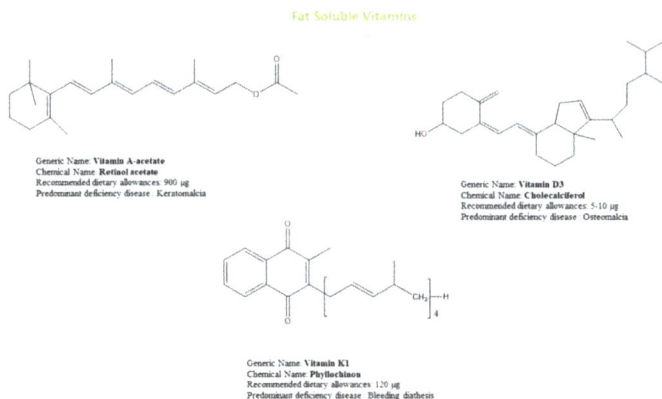

Generic Name: **Vitamin A-acetate**
Chemical Name: **Retinol acetate**
Recommended dietary allowances: 900 µg
Predominant deficiency disease : Keratomalcia

Generic Name: **Vitamin D3**
Chemical Name: **Cholecalciferol**
Recommended dietary allowances: 5-10 µg
Predominant deficiency disease : Osteomalcia

Generic Name: **Vitamin K1**
Chemical Name: **Phyllochinou**
Recommended dietary allowances: 120 µg
Predominant deficiency disease : Bleeding diathesis

Fig.2.0. Names and structures of the Fat Soluble vitamins, their chief scarcity illnesses and the suggested dietary doses

Table 1: Sources of essential vitamins in our day-to-day life.

S. No.	Vitamin Name	Solubility	Sources	References
1	Vitamin-A	Fat	Spinach, carrots, sweet potatoes and red peppers. Yellow fruits, such as mango, papaya and apricots.	1, 34-37
2	Vitamin-B	Water	Meat (especially liver), seafood, poultry, eggs, dairy products, legumes, leafy greens, seeds and fortified foods, such as breakfast cereal and nutritional yeast.	1, 34-37
3	Vitamin-C	Water	Potatoes, green pepper, spinach, lemons, broccoli, strawberries, peas, pears, lime and animal-based sources like pork, chicken, seafood.	1, 34-37
4	Vitamin-D	Fat	Cord liver oil, Salmon, swordfish, Tuna fish, egg yolk, orange juice, and fortified cereals.	1, 34-37
5	Vitamin-E	Fat	Green leafy vegetables, such as kale, spinach, turnip greens, collards, Swiss chard, mustard greens, parsley, romaine, and green leaf lettuce.	1, 34-37
6	Vitamin-K	Fat	Soya bean, cucumbers, asparagus, Apium graveolens dulce, coriander, head lettuce, sauerkraut, kale, brussels sprouts, whole grain, peas.	1, 34-37

5.   Isolation and purgation of vitamin

Since the last decade various chemical and other procedures have been employed to isolate and refine vitamins. The computable studies of vitamins are efficiently done with chromatographic methods. Moreover, chromatographic methods raise the efficiency of the apparatus. It also offers exact outcomes [22]. High performance liquid chromatography (HPLC) is employed to separate the water-soluble and lipid- soluble vitamins but it fails to give satisfactory results. This is due to the disorder of the last peak and the absence of clarity in ultimate isolation in the chromatograph. The following criteria should be followed to isolate and refine vitamins effectively:

1. It must not be a time taking process

2. It should be fast

3. It should be proficient in administering a very low solute-solvent ratio of specimens.

4. It must endure hotness, Ultra-Violet and fountainhead about decomposition.

5. The ultimate outcome should possess a productive clarity.

Chromatographic methods (especially ion-exchange) fulfil all the above-mentioned criterions to isolate and refine vitamin substances [26]. Each and every water soluble vitamin is framed in the ionised state by this procedure. The judicious choice of pH of the eluent can aid in a suitable separation of the multi component arrangement [28].

6.   Ion-exchange chromatography

Ion exchange chromatography is utilised in several isolation and purgation methods of biological molecules [38-41]. The isolation and purgation of the biological molecules by chromatographic methods, especially ion-exchange is based on five different steps: 1. Equilibration, 2. Sorption, 3. Desorption, 4. Elution and 5. Regeneration. Resin is an inevitable part of the chromatographic methods through ion exchange. The main course in ion exchange chromatography is free from minerals, transformation, purgation and congregation [1,22]. In ion exchange chromatography two phases, namely stationary phase and mobile phase bind an ion. Generally, the stationary phase is the hard reticulation polymeric substance. On the other hand, the movable segment is made up of solvents comprising ions. This method particularly separates the charged or ionic molecules. The separation depends on the charges of the molecule. The elementary principle of ion exchange chromatography lies in drawing and attaching among these inversely ionic places of excellently secluded, unsolvable material. It is made up of duets unlike types of ion exchange resins mainly cationic as well as anionic in nature. The positively charged replace resin keeps minus ions and negatively charged replace resin material keeps a positive charge. Ion exchange chromatography excellently separates and purifies the biomolecules for example, nucleotides, proteins and vitamins. For this reason, chromatographic isolation through ion replacement is the best choice for disconnection as well as sanitization of vitamins, polypeptides, nucleoproteins and ionic molecules [1,42-45].

7.   Ion exchange chromatographic isolation and purgation of vitamin K1

Vitamin K1 and K2 constitute the Vitamin K. Phylloquinone or phytyl menaquinone are the other names for Vitamin K1 which is derived from plant sources. Menaquinones, another form of vitamin K2, are created by bacteria and algae. Vitamin K contains 2-methylnaphthalene-1,4-dione rings. It has quite high antihemorrhagic properties. With the help of fatty substances, vitamin K is simply immersed throughout the anatomy. Vitamin K is important for a variety of functions, including bone mineralization, group of cell mineral insertion regulation, blood sugar control by maintaining equity between insulin and glucagon, indicating, pregnancy, tissue mineralization modulation in addition to eicosanoids consumption. Naturally Vitamin K1 is found in the trans-form, while it is found in the cis-form when manufactured synthetically. The trans-form of vitamin K is extremely energetic. To accurately estimate the dietary assessment of a complementary component, it should be extremely desirable for the isolation of vitamin K1 in both trans- and cis forms. Endoplasmic reticulum (ER) stimulated synthesis in liver cells is assisted by vitamin K1. Here the role of vitamin K1 is as a significant cofactor. The transformation of zymogens from vitamin K1 needs blood-coagulating agents and is assisted by vitamin K1 reliant carboxylase [46-48].

8.   Ion exchange chromatographic isolation and purgation of vitamin C

Every bit of citrus fruit supplies a major portion of Ascorbic acid or Vitamin C. Vitamin C is added to the chemically-synthesized berry extract for the enhancement of body immunity. The isolation of vitamin C is generally done with ion exchange chromatographic technique. Here the column is Zipax SAX which is anionic in nature. It is to be noted that the said column should be kept at pH 7 or pH <7. This is because Ascorbic acid contains ionic transformers which are able to oxidise the components in solution [5]. Ascorbic acid is particularly susceptible to oxidation and once oxidised it quickly degrades into the dehydrogenated form which is inactive. For effective isolation of Ascorbic acid, the chromatographic process must be performed at acidic conditions and charged power [28]. Different Ternary ion exchangers are generally employed to determine vitamin C. The first one is a strong cation exchanger e.g., Zipax SCX which is made up of sulphated fluorinated carbon resinous material. This resin is used in an immobilised state in aqueous solvents at pH ranging from one to nine. The drawback is that this resin is not utilised by biotic eluent.

The second resin is a powerful anion exchanger resin e.g., Zipax SAX. This resin is made up of a quartet amino group attached with Zipax resin. The advantage of this resin is that it can be used in an aqueous medium as well. The pitfall of this resin is that it cannot be used in biotic eluents and at elevated heating conditions. The third resin Permaphase AAX overcomes the drawbacks of the first two resins and is easily used in water as well as biotic eluents keeping temperatures over 50 °C. In the pH range of 3–9, it reaches its maximum stability [1,22].

9. Ion exchange chromatographic isolation and purgation of vitamin B1, vitamin B2 and vitamin B6

Owing to their ionic nature, water-soluble vitamins are best isolated and refined by means of ion-exchange chromatography. Vitamin B1 is also called thiamine and its elution power through the column is very poor. Vitamin B1 contains a quartet amino group with a complicated structure. This leads to a problem while elution through firmly retained structural ion exchanger support with cationic nature. Only under extreme ionic strength and pH conditions can thiamine be eluted. A cation-exchange column was used to elute vitamins with amine groups, whereas an anion exchange column was used to separate and purify vitamins with carboxylic or phosphate groups [22]. A cation-exchange column was used to isolate and refine vitamin B2, with the moving stages keeping duo distinct pH levels. The elution of vitamin B2 is carried out through powerful positive ion replace chromatography keeping pH value 1 to 7. When pH value is less than 2, vitamin B2 hangs on to the stationary phase. The anionic exchange chromatography column was also used to separate B2 [1,22]. Pyridoxine, Pyridoxal, and Pyridoxamine are three distinct molecules that make up vitamin B6. The Most familiar type of vitamin B6 is Pyridoxine, which is generally used as a source of nutrition in pharmaceutical products. Zipax SCX positive ion replacer column aids to elute the trio Vitamin B6 molecules at pH 4. Zipax SCX was used to separate Vitamins B2 and B6 at an acidic pH using a phosphate buffer [22,49]. Zeolite resin successfully isolates and refines Vitamin molecules. On the other hand alumina resin is unable to isolate vitamins. Amberlite IR-100 resin was used to successfully separate thiamine and riboflavin. Amberlite IR-4 adsorbs vitamin B1 due to the acidulated nature of the resin material. Thiamine adsorption by Amberlite IR-4 was not as efficient as expected. Vitamins, rather than zeolite compounds, bind robustly with resins containing energetic groups in the separation of Vitamins [50-52]. Hilker and Clifford calculated vitamin B1 from corn and pee specimens by using a 5-mm NH Microlab AX5 column. This column is paired with a phosphate buffer with a pH of roughly 2.85 [55-58].

Conclusion

The important nutrient vitamins play a vital role in the maturation of the human body system. A dozen classes of Vitamins are required for this purpose. The major categories of vitamins are water-soluble and fat-soluble vitamins. Vitamin B and C are soluble in aqueous medium. Vitamin B is subdivided into Vitamin B1, Vitamin B2, Vitamin B3, Vitamin B6. Several disunion methods are employed for the partition and purgation of vitamins. In this context, ion exchange resins are widely used to detach and refine the entire biological molecules together with vitamins. Ion exchange resins are of two types; I. cation exchange resin, II. anion exchange resin. Both the types are successfully utilised for the partition of vitamins.

References

[1] P.S. Kumar, G. J.Joshiba, Separation and Purification of Vitamins: Vitamins B1, B2, B6, C and K1, in: Inamuddin (Eds.), Applications of Ion Exchange Materials in Biomedical Industries, Springer Nature, Switzerland AG, 2019, pp. 177-186. https://doi.org/10.1007/978-3-030-06082-4_9

[2] M.Wenning, T. Hu, C. Li, R. Wu, J. Chen, Y. Shao, J. Liang, Y. Wei, Effective nitrogen removal of wastewater from vitamin B2 production by a potential anammox process, Journal of Water Process Engineering. 37 (2020) 101515. https://doi.org/10.1016/j.jwpe.2020.101515

[3] A. Tucaliuc, A.Cîşlaru, L.Kloetzer, A. C. Blaga, Strain development, substrate utilization, and downstream purification of Vitamin C, Processes. 10 (8) (2022) 1595. https://doi.org/10.3390/pr10081595

[4] G. Homaie, Solmaz, K. A. Almanghadim, M. Mirghafourvand, Effect of vitamins on sexual function: A systematic review, International Journal for Vitamin and Nutrition Research. 1 (1) (2021) 1-10. https://doi.org/10.1024/0300-9831/a000703

[5] T. J. Smith, C. R. Johnson, R. Koshy, S.Y. Hess, U. A. Qureshi, M. L.Mynak, P. R. Fischer, Thiamine deficiency disorders: a clinical perspective, Annals of the New York Academy of Sciences. 1498 (1) (2021) 9-28. https://doi.org/10.1111/nyas.14536

[6] A. F.Valevski, Thiamine (vitamin B1), Journal of Evi. based Integ. Med. 16 (2011) 12-20. https://doi.org/10.1136/ebm1131

[7] K. C. Whitfield, M. W. Bourassa, B.Adamolekun, G. Bergeron, L.Bettendorff, K. H. Brown, L. Cox et al. Thiamine deficiency disorders: diagnosis, prevalence, and a roadmap for global control programs, Annals of the New York Academy of Sciences. 1430 (1) (2018) 13919. https://doi.org/10.1111/nyas.13919

[8] S. Dhir, T.Maya, E. Napoli, C.Giulivi, Neurological, psychiatric, and biochemical aspects of thiamine deficiency in children and adults, Frontiers in psychiatry. (2019) 207. https://doi.org/10.3389/fpsyt.2019.00207

[9] E. P. Odum, V. C. Wakwe, Plasma concentrations of water soluble vitamins in metabolic syndrome subjects, Nigerian Journal of Clinical Practice. 15 (4) (2012) 442-447. https://doi.org/10.4103/1119-3077.104522

[10] M. C. Wood, J. A. Tsiouris, M.Velinov, Recurrent psychiatric manifestations in thiamine-responsive megaloblastic anemia syndrome due to a novel mutation c. 63_71 delACCGCTC in the gene SLC19A2, Psychiatry and Clinical Neurosciences. 68 (6) (2014) 487-487. https://doi.org/10.1111/pcn.12143

[11] T. J. Smith, C. R. Johnson, R. Koshy, S. Y. Hess, U. A. Qureshi, M.L. Mynak, P.R. Fischer, Thiamine deficiency disorders: a clinical perspective, Annals of the New York Academy of Sciences. 1498 (1) (2021) 9-28. https://doi.org/10.1111/nyas.14536

[12] L.Cole, P. R.Kramer, Vitamins and minerals, in: Human physiology, biochemistry and basic medicine, Academic Press, London, 2016, pp. 165-175. https://doi.org/10.1016/B978-0-12-803699-0.00037-2

[13] N. Sanlier, B. B. Gokcen, M. Altuğ, Tea consumption and disease correlations, Trends in Food Science & Technology. 78 (2018) 95-106. https://doi.org/10.1016/j.tifs.2018.05.026

[14] C. S. Silva, C. Moutinho, A. F. d. Vinha, C. Matos, Trace minerals in human health: Iron, zinc, copper, manganese and fluorine, International Journal of Science and Research Methodology. 13 (3) (2019) 57-80.

[15] J. F. Zahra, A. E. L. Moussaoui, M. Bourhia, H. Imtara, H. Saghrouchni, K. Ammor, H. Ouassou et al, Anacyclus pyrethrum var. pyrethrum (L.) and Anacyclus pyrethrum var. depressus (Ball) Maire: Correlation between total phenolic and flavonoid contents with antioxidant and antimicrobial activities of chemically characterized extracts, Plants. 10 (1) (2021) 149. https://doi.org/10.3390/plants10010149

[16] J. Söder, S. Wernersson, J. Dicksved, R. Hagman, J. R. Östman, A. A. Moazzami, K. Höglund, Indication of metabolic inflexibility to food intake in spontaneously overweight Labrador Retriever dogs, BMC Veterinary Research. 15 (1) (2019) 1-11. https://doi.org/10.1186/s12917-019-1845-5

[17] D. O.Kennedy, B vitamins and the brain: mechanisms, dose and efficacy: a review, Nutrients. 8 (2016) 68. https://doi.org/10.3390/nu8020068

[18] D. M. Chu, J. Ma, A. L. Prince, K. M. Antony, M.D. Seferovic, K. M. Aagaard, Maturation of the infant microbiome community structure and function across multiple body sites and in relation to mode of delivery, Nature medicine. 23 (3) (2017) 314-326. https://doi.org/10.1038/nm.4272

[19] S. Salminen, M. C. Collado, A. Endo, C. Hill, S. Lebeer, E. M. Quigley, M. E.Sanders, R. Shamir, J. R. Swann, H. Szajewska, G. Vinderola, The International Scientific Association of Probiotics and Prebiotics (ISAPP) consensus statement on the definition and scope of postbiotics, Nature Reviews Gastroenterology & Hepatology. 18 (9) (2021) 649-667. https://doi.org/10.1038/s41575-021-00440-6

[20] C. A. Calderón-Ospina, M. O. Nava-Mesa, B Vitamins in the nervous system: Current knowledge of the biochemical modes of action and synergies of thiamine, pyridoxine, and cobalamin, CNS Neuroscience & Therapeutics. 26 (1) (2020) 5-13. https://doi.org/10.1111/cns.13207

[21] A. L. Tardy, E. Pouteau, D. Marquez, C. Yilmaz, A. Scholey, Vitamins and minerals for energy, fatigue and cognition: a narrative review of the biochemical and clinical evidence, Nutrients. 12 (1) (2020) 228. https://doi.org/10.3390/nu12010228

[22] R. C.Williams, D. R.Baker, J. A.Schmit, Analysis of water-soluble vitamins by high-speed ion-exchange chromatography, J.Chromatogr. Sci. 11 (1973) 618-624. https://doi.org/10.1093/chromsci/11.12.618

[23] S. P.Arya, M. Mahajan, P. Jain, Non-spectrophotometric methods for the determination of Vitamin C, Analytica Chimica Acta. 417 (1) (2000) 1-14. https://doi.org/10.1016/S0003-2670(00)00909-0

[24] P. W. Washko, R. W. Welch, K. R. Dhariwal, Y.Wang, M. Levine, Ascorbic acid and dehydroascorbic acid analyses in biological samples, Analytical biochemistry. 1 (1992) 1-14. https://doi.org/10.1016/0003-2697(92)90131-P

[25] S. Albala, M. T. Veciana-Nogue, A. Marine, Determination of water-soluble vitamins in infant milk by high-performance liquid chromatography, Journal of Chromatography A. 778 (1-2) (1997) 247-253. https://doi.org/10.1016/S0021-9673(97)00387-7

[26] V.V.Khasanov, K. A.Dychko, G. L.Ryzhova, Analysis of water-soluble vitamins by ion exchange chromatography, Pharm. Chem. J. 32 (1998) 45-46. https://doi.org/10.1007/BF02464175

[27] L. I. Luttseva, L. G. Maslov, Methods of control and standardization of drugs containing water-soluble vitamins :a Review, Pharmaceutical Chemistry Journal. 33 (9) (1999) 490-498. https://doi.org/10.1007/BF02510075

[28] A.Fallon, R. F. GBooth, L. D. Bell, Vitamins, in: R. H. Burdon, P. H. van Knippenberg (Eds.), Applications of HPLC in Biochemistry, Elsevier, 1987, pp. 271-294.

[29] B. Klejdus, J. Petrlová, D. Potěšil, V. Adam, R. Mikelová, J. Vacek, R. Kizek, V. Kubáň, Simultaneous determination of water-and fat-soluble vitamins in pharmaceutical preparations by high-performance liquid chromatography coupled with diode array detection, Analytica Chimica Acta. 520 (1-2) (2004) 57-67. https://doi.org/10.1016/j.aca.2004.02.027

[30] A.G.Vilaplana, D.Villano, J.Marhuenda, D. A.Moreno, C. G.Viguera, Nutraceutical and functional food components, in: C. A.Galanakis (Eds.),Vitamins, Academic Press, Tokyo, 2017, pp. 159-201.

[31] C. K.Markopoulou, K. A.Kagkadis, J. E.Koundourellis, An optimized method for the simultaneous determination of vitamins B1, B6, B12, in multivitamin tablets by high performance liquid chromatography, J. Pharm. Biomed. Anal. 30 (2002) 1403-1410. https://doi.org/10.1016/S0731-7085(02)00456-9

[32] M. Aranda, G. Morlock, Simultaneous determination of riboflavin, pyridoxine, nicotinamide, caffeine and taurine in energy drinks by planar chromatography-multiple detection with confirmation by electrospray ionization mass spectrometry, Journal of Chromatography A. 1131 (1-2) (2006) 253-260. https://doi.org/10.1016/j.chroma.2006.07.018

[33] B. Klejdus, J. Petrlová, D. Potěšil, V. Adam, R. Mikelová, J. Vacek, R. Kizek, V. Kubáň, Simultaneous determination of water-and fat-soluble vitamins in pharmaceutical preparations by high-performance liquid chromatography coupled with diode array detection, Analytica Chimica Acta. 520 (1-2) (2004) 57-67. https://doi.org/10.1016/j.aca.2004.02.027

[34] D.Guggisberg, M. C.Risse, R.Hadorn, Determination of Vitamin B12 in meat products by RP-HPLC after enrichment and purification on an immuno-affinity column, Meat. Sci. 90 (2012) 279-283. https://doi.org/10.1016/j.meatsci.2011.07.009

[35] Y. Zhang, W. E. Zhou, J. Q. Yan, M. Liu, Y. Zhou, X. Shen, Y. L. Ma, X. S. Feng, J. Yang, G. H. Li, A review of the extraction and determination methods of thirteen essential vitamins to the human body: An update from 2010, Molecules. 23 (6) (2018) 1484. https://doi.org/10.3390/molecules23061484

[36] B. Chamlagain, M. Edelmann, S. Kariluoto, V. Ollilainen, V. Piironen, Ultra-high performance liquid chromatographic and mass spectrometric analysis of active vitamin B12 in cells of Propionibacterium and fermented cereal matrices, Food chemistry. 166 (2015) 630-638. https://doi.org/10.1016/j.foodchem.2014.06.068

[37] K. R. Heal, L. T. Carlson, A. H. Devol, E. V. Armbrust, J. W. Moffett, D. A. Stahl, A. E. Ingalls, Determination of four forms of vitamin B12 and other B vitamins in seawater by liquid chromatography/tandem mass spectrometry, Rapid Communications in Mass Spectrometry. 28 (22) (2014) 2398-2404. https://doi.org/10.1002/rcm.7040

[38] F.Dechow, Ion exchange, in: M. Silva (Eds.), Separation and Purification techniques in Biotechnology, Noyes Publications, Park Ridge, NJ, 1989, pp.163-332.

[39] B. Van de Voorde, B. Bueken, J. Denayer, D. De Vos, Adsorptive separation on metal-organic frameworks in the liquid phase, Chemical Society Reviews. 43 (16) (2014) 5766-5788. https://doi.org/10.1039/C4CS00006D

[40] H. H. Tran, F. A. Roddick, J. A. O'Donnell, Comparison of chromatography and desiccant silica gels for the adsorption of metal ions-I. adsorption and kinetics, Water Research. 33 (13) (1999) 2992-3000. https://doi.org/10.1016/S0043-1354(99)00017-2

[41] Z. Ma, R. D. Whitley, N. H. Wang, Pore and surface diffusion in multicomponent adsorption and liquid chromatography systems, AIChE Journal. 42 (5) (1996) 1244-1262. https://doi.org/10.1002/aic.690420507

[42] S.D.Alexandratos, Ion-exchange resins: a retrospective from industrial and engineering chemistry research, Ind. Eng. Chem. Res. 48 (2009) 388-398. https://doi.org/10.1021/ie801242v

[43] S. Sharma, A. Bhattacharya, Drinking water contamination and treatment techniques, Applied water science. 7 (2017) 1043-1067. https://doi.org/10.1007/s13201-016-0455-7

[44] J. Balapanuru, J. X. Yang, S. Xiao, Q. Bao, M. Jahan, L. Polavarapu, J. Wei, Q. H. Xu, K. P. Loh, A graphene oxide-organic dye ionic complex with DNA-sensing and optical-limiting properties, Angewandte Chemie International Edition. 49 (37) (2010) 6549-6553. https://doi.org/10.1002/anie.201001004

[45] M. M. Nasef, O. Güven, Radiation-grafted copolymers for separation and purification purposes: Status, challenges and future directions. Progress in Polymer Science. 37 (12) (2012) 1597-1656. https://doi.org/10.1016/j.progpolymsci.2012.07.004

[46] L. A.Begent, A. P.Hill, G. B.Steventon, A. J.Hutt, C. JPallister, D. C.Cowell, Characterization and purification of the vitamin K1, 2, 3 epoxide reductase system from rat liver, J. Pharm. Pharmacol. 53 (2001) 481-486. https://doi.org/10.1211/0022357011775776

[47] S. Rost, A. Fregin, V. Ivaskevicius, E. Conzelmann, K. Hörtnagel, H. J. Pelz, K. Lappegard, E. Seifried, I. Scharrer, E. G. Tuddenham, C. R. Müller, Mutations in VKORC1 cause warfarin resistance and multiple coagulation factor deficiency type 2, Nature. 427 (2004) 537-541. https://doi.org/10.1038/nature02214

[48] T. Li, C. Y. Chang, D. Y. Jin, P. J. Lin, A. Khvorova, D. W. Stafford, Identification of the gene for vitamin K epoxide reductase, Nature. 427 (2004) 541-544. https://doi.org/10.1038/nature02254

[49] K.Callmer, L.Davies, Separation and determination of vitamin B1, B2, B3 and nicotinamide in commercial vitamin preparations using high performance cation-exchange chromatography, Chromatographia. 7 (1974) 644-650. https://doi.org/10.1007/BF02290508

[50] R. B. Toma, M. M. Tabekhia. High performance liquid chromatographic analysis of B-vitamins in rice and rice products, Journal of Food Science. 44 (1) (1979) 263-265. https://doi.org/10.1111/j.1365-2621.1979.tb10057.x

[51] A. Monir, J. Reusch. "High-performance liquid chromatography of water-soluble vitamins. Part 3. Simultaneous determination of vitamins B 1, B 2, B 6, B 12 and C, nicotinamide and folic acid in capsule preparations by ion-pair reversed-phase high-performance liquid chromatography", Analyst. 112 (7) (1987) 989-991. https://doi.org/10.1039/an9871200989

[52] D.S. Herr, Synthetic ion exchange resins in the separation, recovery, and concentration of thiamine, Ind. Eng. Chem. 37 (1945) 631-634. https://doi.org/10.1021/ie50427a011

[53] S. D. Alexandratos, Ion-exchange resins: a retrospective from industrial and engineering chemistry research, Industrial & Engineering Chemistry Research. 48 (2009) 388-398. https://doi.org/10.1021/ie801242v

[54] K. A. Edwards, N. Tu-Maung, K. Cheng, B. Wang, A. J. Baeumner, C. E. Kraft, Thiamine assays-advances, challenges, and caveats, ChemistryOpen. 6 (2) (2017) 178-191. https://doi.org/10.1002/open.201600160

[55] Lynch PLM, Young IS, Determination of thiamine by high-performance liquid chromatography, J.Chromatogr. A. 881 (2000) 267-284. https://doi.org/10.1016/S0021-9673(00)00089-3

[56] G. F. Ball, Vitamins in foods: analysis, bioavailability, and stability, CRC Press, 2005. https://doi.org/10.1201/9781420026979

[57] R. B. Rucker, J. Zempleni, J. W. Suttie, D. B. McCormick, Handbook of vitamins, CRC Press, 2007. https://doi.org/10.1201/9781420005806

[58] C.J. Blake, Analytical procedures for water-soluble vitamins in foods and dietary supplements: a review, Analytical and bioanalytical chemistry. 389 (2007) 63-76. https://doi.org/10.1007/s00216-007-1309-9

Ion Exchange Resins: Biomedical and Environmental Applications   Materials Research Forum LLC
Materials Research Foundations 137 (2023) 39-54   https://doi.org/10.21741/9781644902219-3

Chapter 3

Application of Ion Exchange Resins in Protein Separation and Purification

Srijita Basumallick

Asutosh College, 92, S.P.Mukherjee Road, Kolkata-700026, India

Abstract

Separation and purification of proteins obtained from natural sources is a really challenging job. This chapter aims to discuss various aspects of separation and purification of proteins by ion exchange chromatographic method. Emphasis has been given on understanding the basic principles and different factors that govern the efficiency and commercial applications of separation of protein obtained by this method.

Keywords

Affinity Chromatography, Immunoaffinity Chromatography, Gel Filtration or Permeation Chromatography, Buffer Solution, Cation and Anion Exchange Chromatography, Donnan Equilibrium

Contents

Application of Ion Exchange Resins in Protein Separation and Purification..39

1.   Basic principle of protein separation and purification by chromatographic method 40

2.   Chromatographic methods of protein purification............................41

     2.1   Gel filtration or permeation chromatography.....................................41

     2.2   Affinity chromatography 42

     2.3   Immuno affinity chromatography...43

     2.4   Metal chelate chromatography 43

     2.5   Other Chromatographic techniques 43

3.   Principle of separation of proteins by ion exchange
chromatography..44

4.   Strong and weak ion exchange resin.....................................44

5.   Choice of buffer...45

6.   Experimental procedure of ion exchange resin 46
       6.1   Equilibration 47
       6.2   Sample Application and Wash 47
       6.3   Elution..47
       6.4   Regeneration 47

7.   Morphology of ion exchange resin 47
       7.1   Capacity of ion exchange resin......................................47
       7.2   Stability...48
       7.3   Cross linking of resins 48
       7.4   Donnan equilibrium 48

8.   Parameters for optimisation of ion exchange methods.......49
       8.1   Resolution 49
       8.2   Efficiency..50
       8.3   Selectivity 50

Summary...51

References 51

1.   Basic principle of protein separation and purification by chromatographic method

The basic principle of chromatographic separation and purification of protein is quite simple. In fact, the surface charge of the protein is used as a basic distinguishing factor in ion exchange chromatography. Whereas size of the protein is used in size exclusion chromatography [1]. It is also called gel filtration chromatography. Hydrophobicity of proteins is used in hydrophobic interaction chromatography [2], which is also called reversed phase chromatography. Finally, in affinity chromatography [3] bio-recognition of protein by stationary phase is used as the underlying principle. Actually, a ligand specific binding [4] works as an underlying mechanism. Among all other processes, ion exchange

Ion Exchange Resins: Biomedical and Environmental Applications   Materials Research Forum LLC
Materials Research Foundations 137 (2023) 39-54   https://doi.org/10.21741/9781644902219-3

chromatography is important because it can be easily fine-tuned with the help of buffer pH. For example, in a buffer of pH 4, the surface charges of proteins in a mixture in general vary to a large extent as depicted in figure 1.

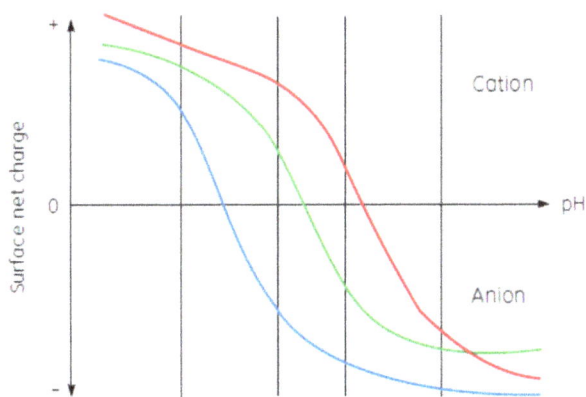

Figure 1. Theoretical protein titration curve showing different surface charge at pH x.

2. Chromatographic methods of protein purification

2.1 Gel filtration or permeation chromatography

This type of chromatography uses molecular shape and size of two separate protein molecules. It can also be called gel exclusion or size exclusion, molecular sieve chromatography or simply gel chromatography. Smaller molecules enter matrix pore where bigger molecules are excluded. Bigger molecules come fast [5] out of these columns whereas smaller come later. To avoid diffusion, the diameter of chromatographic columns are kept small with a long height. Also, in gel chromatography gravity filtration is replaced by a special pump used to maintain the elution process at constant velocity. Now for bigger proteins or totally excluded particles that don't enter the pore, the elution volume is taken to be void [6] equal to volume V_0 (= total physical column volume (V_t) -volume of stationary phase (V_s)). Whereas for smaller and totally included particles, elution volume is V_t (total physical volume of column including pore volume or volume of stationary phase V_s + void volume V_0). Apart from that we can use the equation:

$$V_e = V_0 + K_{av} (V_t - V_0)$$

Materials Research Forum LLC
https://doi.org/10.21741/9781644902219-3

where, $0<K_{av}<1$.

Here K_{av} is the portion of pores available to the molecules and V_e = elution volume. Here it is worth mentioning that for HPLC the column length is low but due to small beads tight packing the number of plates is high [7].

2.2 Affinity chromatography

Affinity chromatography [8] works through some highly selective interactions like inhibitor-receptor hormone interaction, avidin-biotin affinity, simple antibody-antigen interactions etc. It produces highly recoverable proteins. It exploits these specific interactions between protein and any of these specific ligands (bound to column matrix, like agarose or dextran through 6-8 carbon spacer arm or direct bond formation) that binds to the protein reversibly. Ligands and protein interactions [9] are purely electrostatic or Van Der Waals type [10] or hydrogen bonding. To make it reversible, the exchange process can be done specifically by competitive ligand binding [11] or non specifically by pH or ionic strength [12] adjustment. Matrix pore size [13] as well as spacer arm length [14] help remove steric hindrance. The latter one helps in binding big proteins without steric hindrance from matrix polymers. This spacer might contain other functional groups like amide, keto, etc [15]. The spacer molecule should be chemically stable and should be specific to the target protein. Special buffers help maintain the pH for strong ligand protein bonding [16,17]. In case of competitive ligand binding [18], same or similar ligands that of stationary phase are frequently used for elution. Ligand coupling is another important point where we need to bind ligand covalently to a spacer attached to a stationary phase, without losing its affinity to protein. This ligand can be mono specific i.e specific for only one protein or it can be specific for a group of proteins. But ligands should bind reversibly to protein and be stable under elution conditions. It should be resistant to proteolysis and denaturation caused by elution. To overcome stereo selectivity in the affinity chromatography, big ligands are used for binding more proteins. Excessively large ligands are prone to denaturation and degradation and can also become non-specific sometimes. Matrix polymer is activated through coupling reactions and grafted with ligands. Concentrations of coupling reagents are important for the coupling reactions like cyanogen bromide activated agarose, 6-amino hexanoic acid and 1,6- di amino hexane-agarose, epoxy activated agarose, thiopropyl agarose, carbonyldiimidazole activated agarose, amino ethyl and hydrazide activated agarose. Similarly, concentrations of ligands on matrix are also important for the coupling reactions. There are other types of affinity chromatographic techniques that are known, for example, lectin affinity chromatography: Lectin can bind to carbohydrates and thus glycoproteins. Lectins can be monomeric or trimeric with identical subunits or different subunits. Lectin with identical subunits recognises single glycoprotein, these are useful in purifying membrane receptor glycol-proteins. Elution of bound protein is done by several ways like using mono-saccharides in a competitive way. Borate buffers often form complexes with glyco-protein. Changing pH or using ethylene glycol helps reduce hydrophobic interactions among proteins. Sometimes ionic strength is allowed to vary to reduce hydrophobic interaction as in the case of lectin chromatography.

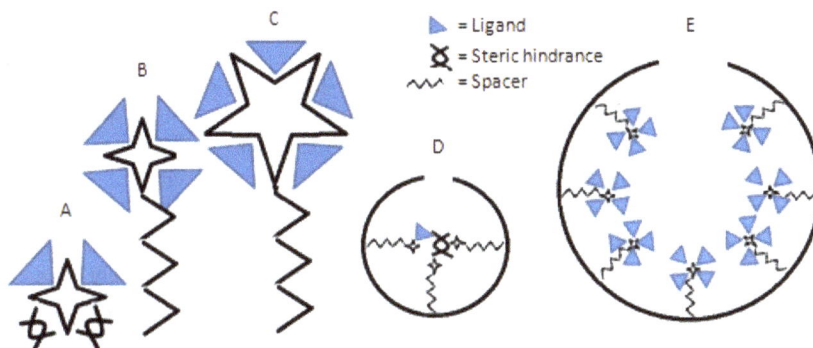

Figure 2. Schematic representation showing A. Steric hindrance in case of ligand directly bound to matrix. B. Spacer-reduced steric hindrance. C. Big ligand binding more protein. D. Small pore causing steric hindrance. E. Big pore causing least steric hindrance.

2.3   Immuno affinity chromatography

Immuno affinity is specifically based on antibody-antigen interactions. Here the antibody is used as an immobilized ligand. Mono chloral antibodies link to agarose matrix. Proteins of any virology origin can be purified varying ionic strength of eluent or using urea or guanidine hydrochloride or lowering the pH.

2.4   Metal chelate chromatography

In this technique, immobilized cations like copper or zinc or mercury are used. Here the imidazole group of histidine molecules, cysteine or indole groups of tryptophan bind to these metal ions. Lowering the pH or using complexing agents like EDTA help elution. Nickel ion is used to purify histidine tagged or polyhistidine tagged proteins. Zinc columns can be utilised for isolation of human interferon.

2.5   Other Chromatographic techniques

There may be some other chromatographic techniques like dye ligand chromatography that can separate DNA binding proteins. In covalent bond forming chromatographic techniques, thiol containing proteins with disulphide 2-pyridyl group are often immobilized on matrix. The best part is that the quantification of protein binding is done measuring absorption at 343 nm for thione form (another form of 2-thiopyridine), released due to bond breaking.

3. Principle of separation of proteins by ion exchange chromatography

Proteins from natural sources often occur as mixed forms of almost similar structures. Thus, their separation is a difficult task. The principle of separation of proteins by ion exchange chromatography depends on the attraction between the oppositely charged stationary phase and analyte. The stationary phase is called ion exchanger and protein mixture being the analyte. There are mainly two types of ion exchangers 1. cation exchange resin and 2. anion exchange resin. Cation exchange resins are positively charged resins and they attract negatively charged ions on their surface. Basically, higher strength negatively charged molecules can replace lower strength negatively charged molecules on the surface. This action leads to ion exchange. In cation exchange resins, there are negatively charged functional groups on the surface that attract cations. Another classification based on acidity of resin says that basic functional groups on ion exchange resin are anion exchange resins and acidic fictional groups on ion exchange resin are cation exchange resins. DEAE (diethylaminoethyl) [19] cellulose is an example of anion exchange resins. As we have just discussed, in an anion exchanger there is a positive charge on its surface where a proton attached to the ammonium ion contributes to the positive charge. That is why this ion exchanger is also called the basic ion exchanger. Perfect example for cation exchange resins is CM (carboxymethyl) cellulose. In this the negative charge is because of the ionisation of the carboxylic acid so that COOH is dissociated into COO-. Again negative charge on the carboxylate ion acts as an cationic exchanger. Ion exchange chromatography (IEX) follows deamidation and succinimide formation. Ion exchange chromatography depends on the net charge of the protein.

4. Strong and weak ion exchange resin

Ion exchange resins can further be classified as strong and weak ion exchangers [20]. Strong ion exchangers are those who have functional groups that are totally ionised at all working pH values. Sulfonate (SO_3^-) [21] ions and quaternary ammonium ($-NR_3^+$) [22] ions are the best examples because both sulphonate (SO_3^-) ions and quaternary ammonium ($-NR_3^+$) ions are in negatively and positively ionised state in a wide range of pH value. Weak ion exchangers are those with functional groups that can ionise within a small window of buffer pH. At a very high or low buffer pH, functional group dissociation following degradation takes place on the surface of the resin packed column. Dimethyl ammonium and carboxylate groups as functional ions on the resin surface maintain their ionic state inside a narrow pH range. We need stability of test analytes to be seen inside a particular pH range [23] (of interest) on that resin's surface for best choices to be made in choosing ion exchange resins. If the protein is stable in a narrow pH range then weak ion exchangers are selected. If a protein is stable below its isoelectric point that is in its cationic form, cation exchange resin is selected. But if the protein is stable above its isoelectric point that means it is stable in its anionic form and anion exchanger is needed in separating the protein analytes. If a protein is stable in a wide range of pH value then either cationic or anionic exchanger may be used during protein separation. Strong exchangers are used for separating weak electrolytes as weak analytes ionise only in the very low or very high pH range. Strong exchangers work within a wide gap of pH range. Thus, weak analytes can be deprotonated or protonated in a very high or low pH window to fit and get purification done. But for strong analytes, both strong and weak exchangers can be used. But practically a weak exchanger is selected that works near its physiological pH. That helps in two ways: first protein is always stable near physiological pH and the second being all the weakly charged impurities near physiological pH bind weakly to the ion exchange resin surface making it easy to elute and get rid of these impurities. Differences and characteristics of weak and strong cation and anion exchange resins are shown in Table 1 [24].

5.   Choice of buffer

Now based on previously discussed ideas, the pH of the buffer can be chosen [25]. The buffer used in an ion exchange chromatography is chosen in such a way that its pH is 1 unit separate from the protein's isoelectric point. Buffer used for the cation exchange resin method should not contain any strong acid or alkali. Such buffers are tris buffer, pyridine, alkyl amine base, these are anionic buffers. Whereas in anion exchange chromatography perfect buffers are acetate, barbiturate and phosphate and they are cationic buffers as no anions present. Initial buffer pH and ionic strength should allow the binding of analytes to the exchanger. The lowest ionic strength buffer should be used initially to allow minimum binding of contaminants or ions of buffer. Also buffer pH should be maintained in such a way that the analyte binds fully to the exchanger. All amino acids and any covalently attached modifications in protein govern net charge on proteins. The net charge of a protein depends on buffer pH. A buffer with pH = $pI \pm 1$ is usually sufficient for protein binding.

Table 1. Comparison of different properties of anion and cation exchange resins for protein separation.

Properties	Anion Exchanger	Cation Exchanger
Charge on stationary phase.	Positively charged groups on stationary state.	Negatively charged groups on stationary state.
Molecules that it exchanges with:	Attracts negatively charged ions	Attracts positively charged ions
Alternate name:	It is also called basic ion exchanger as it binds with H+ ion with anionic or basic functional groups.	It is also called acidic ion exchanger as it releases H+ ions from acidic functional groups.
Subdivision	This is further divided in two different sub groups a. Strong anion exchanger and b. weak anion exchanger.	This is also further divided in two different sub groups a. Strong cation exchanger and b. weak cation exchanger.
Functional groups	a. Quaternary ammonium (strong anion exchanger) b. Diethylaminoethyl (DEAE) (weak anion exchanger)	a. Sulfopropyl (SP) (strong cation exchanger) b. Carboxymethyl (CM) (weak cation exchanger)
Commercial resin	a. Amberlite, Dowex 1X2 (strong anion exchanger) b. DEAD cellulose (weak anion exchanger)	a. SP Sephadex (strong cationic exchanger) b. CM Cellulose (weak cationic exchanger)
Working range of pH	a. pH 2-12 (strong anion exchanger) b. pH 2-9 (weak anion exchanger)	a. pH 4-13 (strong cation exchanger) b. pH 6-10 (weak cation exchanger)

6. Experimental procedure of ion exchange resin

There are two types of elutions, gradient elution and isocratic elution. In case of anion exchange, the pH gradient is decreased for a gradient elution. At low pH, anion gets protonated creating loosely bound molecules on an anion binding site of anion exchange resin. Thus anion exchange increases during elution. For cation exchanger, pH gradient [26] is increased as it loses proton or cationic form and becomes unstable and breaks its bonding with the resins matrix and easily comes out . In both cation and anion exchangers, the ionic strength of the buffer needs to be increased with either increase or decrease in the pH gradient. There are four steps in ion exchange chromatography; first equilibration, second sample application, third elution and fourth regeneration of the resin [27].

6.1 Equilibration

In the equilibration step, the counter ion present in the buffer comes in equilibrium with the resin surface where the analyte can bind after replacing these ions. The pH and ionic strength is important in a buffer to start with. This is to ensure that only proteins of interest bind to the medium.

6.2 Sample Application and Wash

Here the analyte is loaded to the column with as much high concentration as possible to improve the efficiency and resolution of separation. Here protein needs to bind reversibly and all unbound material present needs to be washed out from the column.

6.3 Elution

The proteins that bind weakly to column material (lowest net charge at starting buffer) will be the first ones eluted from the column. Similarly, the proteins that bind strongly to column material (highest charge at that elution pH) will be most strongly retained and will be eluted last. Overall ionic strength governs charge on the protein (to be purified). Also, solvent interactions and ion dipole interactions are important that actually loosen proteins from the stationary surface as well.

The rate of elution plays an important role. Higher capacity of a protein (explained later) on a particular ion exchange resin will be the determining factor for adjusting the speed. Also smaller ions can equilibrate faster between solvent and stationary phase. This is due to the higher diffusion coefficient of small ions compared to larger ions that equilibrate slowly between solvent and stationary phase. Thus the size of analyte ion needs to be taken care before adjusting speed as well. Also unstable molecules need faster filtration and elution. Denser crosslink slows diffusion needs to be accounted for here as well.

6.4 Regeneration

Finally, the column is washed with a salt solution containing counter ions. High ionic strength buffer regenerates the column and removes any molecules from the previous run. The column is equilibrated with the start buffer for the next run.

7. Morphology of ion exchange resin

7.1 Capacity of ion exchange resin

Capacity of ion exchange resins simply means the amount of ions that can be added to a fixed amount of resin. There are three types of capacities: 1. total capacity is the number of charged functional groups present in one gram of dry ion exchanger or in one ml of gel.

2. available capacity, which signifies the actual amount of protein that can bind to an ion exchange under experimental conditions. 3. dynamic capacity, which signifies the actual amount of protein that can bind to an ion exchanger under dynamic elution of solvent. Bead size and polymer nature play an important role in resin chromatography. Resins can either be natural or synthetic. Some examples of plant resins are amber, balm of Gilead, balsam, Canada balsam. Some examples of synthetic resins are polyethylene, polyvinyl chloride (PVC) [28], acrylonitrile butadiene styrene (ABS) [29] etc.

7.2   Stability

Usually ion exchange resins should be quite stable even at high temperatures and under harsh conditions. But at high temperatures, the buffer has significant changes in its pKa values.

7.3   Cross linking of resins

Another very important point of resins is that the extent of cross-linking determines the stability and the rigidity as well as porosity of the resin. Lightly cross-linked resins establish rapid equilibrium of solute inside and outside of the particle. But lightly cross-linked resins have higher hydration which leads to a decrease in analyte concentration as well as the selectivity. On the other hand, heavily cross-linked resins have less swelling or hydration [30] and higher exchange capacity and selectivity but of course a longer equilibration time. Basically pore size in a resin is directly related with flow rate. In the presence of an analyte, the ions inside the resin and the ions outside (that is in the solvent) establish an equilibrium known as Donnan equilibrium.

7.4   Donnan equilibrium

In the case of Donnan equilibrium, selective permeability of ions as well as molecules has been observed. This might be by (1). Sterically hindered protein molecules as well as other big molecules (2). Fixed ions on resin surface (3). Hydrogen bonded ions not allowed to exchange between solvent pools inside the pores of the resin bead and solvent pools outside the resin bead. In spite of the charged groups present on resin's pore surface, electro neutrality as well as chemical potential is balanced between two sides with a net charge separation between pores of resin and its surface. This generates Donnan potential. Due to these extra charges separation together with solvated solvent, net volume of electrolyte is higher inside the pores. These cause expansion in the flexible matrix of resin and internal osmotic pressure is increased. The internal osmotic pressure (π) may be calculated by [31]

$$\pi = \frac{RT}{v_0} \, Ln \left(\frac{(x_0)_0}{(x_0)_i} \right)$$

where $(x_0)_0$ is the mole fraction of water in 0 phase where as $(x_0)_i$ is mole fraction of water in i phase and V_0 is molar volume of water.

For exact ion exchange experiments, the expression gets complicated with introduction of mole fractions of ions as [14],

$$\frac{RT}{V_0} Ln\left[\left(\frac{x_1}{x_2}\right)_i \left(\frac{x_2}{x_1}\right)_0 \left(\frac{(x_0)_0}{(x_0)_i}\right)^{(f-g)}\right] = \pi \left(v_2 - v_1 - (f-g)v_0\right)$$

Where the terms with subscription 0, 1 and 2 refer to those for the solvent, cation A^+ and cation B^+ respectively and V0 denotes molar volume of water. Here A^+ and B^+ are two cations moving from 0 to i phase and i to 0 phase respectively as well as "f" moles of solvent moving from 0 to i phase together with A^+ cation and "g" moles of solvent moving from i phase to 0 phase together with B^+ cation. This inflation near resin's surface helps in diluting the analyte solution as well as changing its activity value. In an actual non-ideal system activity, coefficient gamma is present in the above equation. Another important aspect is that the movable groups and fixed groups can interact with each other with the formation of ion pairs, however, a complex formation may also take place.

8. Parameters for optimisation of ion exchange methods

8.1 Resolution

Resolution of an ion exchange chromatography is determined from both degree of separation between the eluted peaks in terms of the volume of eluent and the ability to produce narrow symmetrical peaks. Also resolution depends on the amount of the analyte used in the beginning. Specially smaller sized particles are used for better resolution. Again separation efficiency as well as peak width depends on column packing and rate of flow of eluent and nature of binding to resin. Resolution is given by,

$$R = \frac{2 \times (V_{R2} - V_{R1})}{W_{b1} + W_{b2}}$$

where V_{R2} and V_{R1} are exact position in terms of eluent volume, Wb1 and Wb2 are width of the base of the peaks.

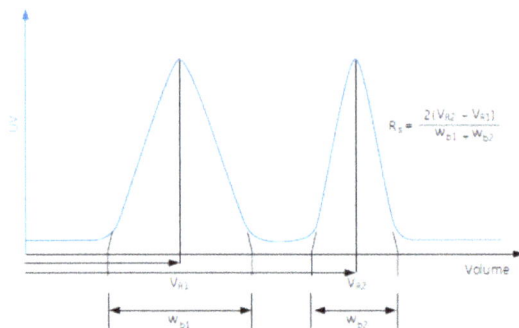

Figure 3. Determination of resolution of chromatographic experiment.

8.2 Efficiency

Efficiency of the column is stated in terms of a theoretical plate which is simply a ratio of height of bed and the height of theoretical plate. The number of theoretical plates is also related by an equation,

$$N = 5.54 \times \left(\frac{V_{R1}}{W_{b1}}\right)^2$$

Where, V_{R1} is equal to net volume of eluded solvent from the start of sample application (i.e V_{R1} for the first eluent as shown in Figure 3) to the peak maximum and W_{b1} equal to the width of the base of the peaks.

The broadening of the peak happens due to longitudinal diffusion of the analyte. This can be controlled by best-packed columns which are in fact not too tightly or not too loosely packed with no air bubbles inside the column. These defects lead to a leakage or uneven passage of analytes with the decrease in efficiency. Lower the particle size, narrower will be the inside channels with increased efficiency.

8.3 Selectivity

Another important point is the selectivity. Selectivity is dependent on the functional groups on matrix, experimental conditions such as pH ionic strength and elution condition. Although concentrated sample loading is prescribed for a higher efficiency, viscosity of highly concentrated samples affects in a bad way in presence of small particles having tight packing. Thus diluted sample loading is a better option.

Summary

In this chapter we have highlighted the basic principles of separation and purification of proteins specifically using the ion exchange chromatography. Different factors governing the efficiency of separation like pH of the buffer solution, nature of ion exchange resin, and the nature of eluting solvent have been discussed. Methodical expressions for calculation of efficiency of resins have also been discussed.

References

[1] R.R. Burgess, A brief practical review of size exclusion chromatography: Rules of thumb, limitations, and troubleshooting, Protein Expr. Purif. 150 (2018) 81-85. https://doi.org/10.1016/j.pep.2018.05.007

[2] S. Fekete, J.L. Veuthey, A. Beck, D. Guillarme, Hydrophobic interaction chromatography for the characterization of monoclonal antibodies and related products, J. Pharm. Biomed. 130 (2016) 3-18. https://doi.org/10.1016/j.jpba.2016.04.004

[3] E.L. Rodriguez, S. Poddar, S. Iftekhar, K. Suh, A.G. Woolfork, S. Ovbude, A. Pekarek, M. Walters, S. Lott, D.S. Hage, Affinity chromatography: A review of trends and developments over the past 50 years, J. Chromatogr. B. 1157 (2020) 122332. https://doi.org/10.1016/j.jchromb.2020.122332

[4] D. Novick, M. Rubinstein, Ligand affinity chromatography, an indispensable method for the purification of soluble cytokine receptors and binding proteins, Methods Mol. Biol. 820 (2012) 195-214. https://doi.org/10.1007/978-1-61779-439-1_12

[5] S.G. Prapulla, N.G. Karanth, FERMENTATION (INDUSTRIAL) | Recovery of Metabolites, in: C.A. Batt, M.L. Tortorello (Eds.), Encyclopedia of Food Microbiology (Second Edition), Academic Press, Oxford, 2014, pp. 822-833. https://doi.org/10.1016/B978-0-12-384730-0.00109-9

[6] C. Ó'Fágáin, P.M. Cummins, B.F. O'Connor, Gel-Filtration Chromatography, Methods Mol. Biol.1485 (2017) 15-25. https://doi.org/10.1007/978-1-4939-6412-3_2

[7] E. Stauffer, J.A. Dolan, R. Newman, Gas Chromatography and Gas Chromatography-Mass Spectrometry, in: E. Stauffer, J.A. Dolan, R. Newman (Eds.), Fire Debris Analysis, Academic Press, Burlington, 2008, pp. 235-293. https://doi.org/10.1016/B978-012663971-1.50012-9

[8] D.S. Hage, Affinity Chromatography: A Review of Clinical Applications, Clin. Chem. 45 (1999) 593-615. https://doi.org/10.1093/clinchem/45.5.593

[9] A.D. Attie, R.T. Raines, Analysis of Receptor-Ligand Interactions, J. Chem. Educ.72 (1995) 119-124. https://doi.org/10.1021/ed072p119

[10] J. Staahlberg, B. Joensson, C. Horvath, Combined effect of coulombic and van der Waals interactions in the chromatography of proteins, Anal. Chem. 64 (1992) 3118-3124. https://doi.org/10.1021/ac00048a009

[11] Magdeldin, S., & Moser, A., Affinity Chromatography: Principles and Applications. In (Ed.), Affinity Chromatography. Intech (2012). https://doi.org/10.5772/39087

[12] E. Sahin, A.O. Grillo, M.D. Perkins, C.J. Roberts, Comparative effects of pH and ionic strength on protein-protein interactions, unfolding, and aggregation for IgG1 antibodies, J. Pharm. Sci. 99 (2010) 4830-4848. https://doi.org/10.1002/jps.22198

[13] M. Sorci, G. Belfort, Insulin Oligomers: Detection, Characterization and Quantification Using Different Analytical Methods, in: V.N. Uversky, Y.L. Lyubchenko (Eds.), Bio-nanoimaging, Academic Press, Boston, 2014, pp. 233-245. https://doi.org/10.1016/B978-0-12-394431-3.00021-3

[14] P. DePhillips, I. Lagerlund, J. Färenmark, A. Lenhoff, Effect of Spacer Arm Length on Protein Retention on a Strong Cation Exchange Adsorbent, J. Anal. Chem.76 (2004) 5816-5822. https://doi.org/10.1021/ac049462b

[15] C. Yu, E.J. Novitsky, N.W. Cheng, S.D. Rychnovsky, Exploring Spacer Arm Structures for Designs of Asymmetric Sulfoxide-Containing MS-Cleavable Cross-Linkers, Anal. Chem. 92 (2020) 6026-6033. https://doi.org/10.1021/acs.analchem.0c00298

[16] X. Du, Y. Li, Y.L. Xia, S.M. Ai, J. Liang, P. Sang, X.L. Ji, S.Q. Liu, Insights into Protein-Ligand Interactions: Mechanisms, Models, and Methods, Int J Mol Sci. 17 (2016) 1-34. https://doi.org/10.3390/ijms17020144

[17] A.V. Onufriev, E. Alexov, Protonation and pK changes in protein-ligand binding, Q. Rev. Biophys. 46 (2013) 181-209. https://doi.org/10.1017/S0033583513000024

[18] D.A. Annis, N. Nazef, C.C. Chuang, M.P. Scott, H.M. Nash, A general technique to rank protein-ligand binding affinities and determine allosteric versus direct binding site competition in compound mixtures, J Am Chem Soc. 126 (2004) 15495-15503. https://doi.org/10.1021/ja048365x

[19] H. SchÄGger, Techniques and Basic Operations in Membrane Protein Purification, in: C. Hunte, G. Von Jagow, H. SchÄGger (Eds.) Membrane Protein Purification and

Crystallization (Second Edition), Academic Press, San Diego, 2003, pp. 19-53. https://doi.org/10.1016/B978-012361776-7/50003-6

[20] A. Staby, J.H. Jacobsen, R.G. Hansen, U.K. Bruus, I.H. Jensen, Comparison of chromatographic ion-exchange resins: V. Strong and weak cation-exchange resins, J. Chromatogr. A 1118 (2006) 168-179. https://doi.org/10.1016/j.chroma.2006.03.116

[21] W.H. Höll, WATER TREATMENT | Anion Exchangers: Ion Exchange, in: I.D. Wilson (Ed.) Encyclopedia of Separation Science, Academic Press, Oxford, 2000, pp. 4477-4484. https://doi.org/10.1016/B0-12-226770-2/04241-1

[22] A.V. Zatirakha, A.D. Smolenkov, A.V. Pirogov, P.N. Nesterenko, O.A. Shpigun, Preparation and characterisation of anion exchangers with dihydroxy-containing alkyl substitutes in the quaternary ammonium functional groups, J. Chromatogr. A. 1323 (2014) 104-114. https://doi.org/10.1016/j.chroma.2013.11.013

[23] G.H. Luttrell, C. More, C.T. Kenner, Effect of pH and ionic strength on ion exchange and chelating properties of an iminodiacetate ion exchange resin with alkaline earth ions, J. Anal. Chem. 43 (1971) 1370-1375. https://doi.org/10.1021/ac60305a048

[24] P.M. Cummins, O. Dowling, B.F. O'Connor, Ion-exchange chromatography: basic principles and application to the partial purification of soluble mammalian prolyl oligopeptidase, Methods Mol. Biol. 681 (2011) 215-228. https://doi.org/10.1007/978-1-60761-913-0_12

[25] D.D. Clark, D.J. Edwards, Virtual protein purification: A simple exercise to introduce ph as a parameter that affects ion exchange chromatography, Biochem Mol Biol Educ. 46 (2018) 91-97. https://doi.org/10.1002/bmb.21082

[26] T. Ahamed, B.K. Nfor, P.D. Verhaert, G.W. van Dedem, L.A. van der Wielen, M.H. Eppink, E.J. van de Sandt, M. Ottens, pH-gradient ion-exchange chromatography: an analytical tool for design and optimization of protein separations, J. Chromatogr. A. 1164 (2007) 181-188. https://doi.org/10.1016/j.chroma.2007.07.010

[27] M. Kosanović, B. Milutinović, S. Goč, N. Mitić, M. Janković, Ion-exchange chromatography purification of extracellular vesicles, Biotechnol. J. 63 (2017) 65-71. https://doi.org/10.2144/000114575

[28] M.V. Srikanth, S.A. Sunil, N.S. Rao, M.U. Uhumwangho, K.V. Ramana Murthy, Ion-Exchange Resins as Controlled Drug Delivery Carriers, J. Sci. Res. 2 (2010) 597. https://doi.org/10.3329/jsr.v2i3.4991

[29] S.M. Hosseini, S.S. Madaeni, A. Reza, Preparation and characterization of ABS/HIPS heterogeneous cation exchange membranes with various blend ratios of polymer binder, J. Membr. Sci. 351 (2010) 178-188. https://doi.org/10.1016/j.memsci.2010.01.045

[30] T. Bruch, H. Graalfs, L. Jacob, C. Frech, Influence of surface modification on protein retention in ion-exchange chromatography. Evaluation using different retention models, J. Chromatogr. A. 1216 (2009) 919-926. https://doi.org/10.1016/j.chroma.2008.12.008

[31] H.P. Gregor, Gibbs-Donnan Equilibria in Ion Exchange Resin Systems, J. Am. Chem. Soc. 73 (1951) 642-650. https://doi.org/10.1021/ja01146a042

Chapter 4

Ion Exchange Resins for Selective Separation of Toxic Metals

Arun Kumar Pramanik[1], Nirmala Tamang[2], Abhik Chatterjee[3*], Ajaya Bhattarai[2*] and Bidyut Saha[4*]

[1]Assistant Chemist, Chemical Laboratory, Damodar Valley Corporation (DVC), India

[2]Department of Chemistry, Mahendra Morang Adarsh Multiple Campus, Tribhuvan University, Biratnagar, Nepal

[3]Department of Chemistry, Raiganj University, Raiganj, Uttar Dinajpur, 733134, WB, India

[4]Department of Chemistry, The University of Burdwan, Burdwan-713104, WB, India

abhikchemistry@gmail.com* (AC); ajaya.bhattarai@mmamc.tu.edu.np* (AB) and bsaha@chem.buruniv.ac.in* (BS)

Abstract

The proper resin selection can make ion exchange a cost-effective and effective pollution control method. Toxic ions in drinking water can be exchanged for other ions via solid ion exchange resin. Nowadays metals are common contaminants in surface water, groundwater, industrial wastewater and other effluent from various sources in the world. Such metals openly challenge and seriously threaten the environment and all living beings. In present time, huge techniques and instruments are developed to separate and filter the metals coming from different sources.

Keywords

Ion Exchange Resin, Toxic Metal, Effective Pollution Control Method

Contents

Ion Exchange Resins for Selective Separation of Toxic Metals....................**55**

1.   Introduction..**56**

2.   Ion exchange resins (IERs) **57**

3.   Type of IERs..**57**

4.   Synthesis of IERs 59

5.   Uses of IERs..60

6.   Activity of IERs..60

7.   Properties of IERs..61

        7.1   IE capacity of resin 61

        7.2   Water retention capacity of ion exchange resin 61

        7.3   Density of ion exchange resin 62

        7.4   Surface area of ion exchange resin..62

        7.5   Regeneration of ion exchange resin 62

8.   Selectivity of IERs..62

9.   Toxic metals..64

10.   Selective separation of toxic metals...65

11.   Modern ion exchange separation method in industry and its future
        prospects..70

Conclusion...71

References 71

1.   Introduction

Metals play an important role in domestic, medical, agriculture, multiple industrial and various technological applications and this leads to them being widespread in nature. The toxic nature, bio-accumulative properties, and persistence of metals in the environment raise concerns about the potential health, environmental, and living body effects of metals. Being persistent pollutants, metals accumulate in nature and consequently contaminate drinking water and the food chains. There are several factors on which metal toxicity depends such as concentration of the metals, age, gender, nutritional status of exposed individuals and chemical species and method of exposure etc. [1]. Some metals like arsenic, mercury, lead and cadmium etc. are renowned as well as possess community health significance because of their high degree of toxicity at low exposure level [2]. But not only human beings but also all animals and plants suffer from the exposure of toxic metals which has increased tremendously as an outcome of an exponential increase of their applications in agriculture, industries, domestic uses and technologies. So there has been a rising environmental and worldwide public health awareness related with the pollution of aquatic and terrestrial ecosystems by the metals. Therefore, a comprehensive study of the removal or purification or separation of the hazardous metals should be taken to minimize the impact of these metals on environment and human health.

One simple, easy and low cost process to separate toxic metals from different mediums or mixtures or solutions or effluents is the ion exchange process by ion exchange resin [3]. By transferring ions between media, ions are transferred from one to the other. Because it is usually a simple reversible reaction, an ion exchanger can usually be reused numerous times. Usually the most common, scientifically efficient and selective ion exchanger is an ion exchange resin.

2.   Ion exchange resins (IERs)

In the process of exchanging different ions, the IERs act as a medium for various organic polymers. Usually, it is in the form of solid micro beads (0.25–1.43 mm radius), yellowish or white, insoluble matrix and composed by organic substrate [4]. It has an active group along with a synthetic functional organic polymer material. It can be obtained by introducing a polymer (cross linked or copolymer) into ion exchange groups and they are attached by covalent bonding. In the resin molecules usually donor atoms are nitrogen and oxygen. For the purification of aqueous solution or mixture, modern ion exchange methods arrest dissolved ions. An IER's most advantageous properties include its capacity to be used in a broad pH range, its insoluble nature in a great deal of aqueous and organic solutions, and its ability to maintain high temperatures. IERs frequently contain polystyrene crosslinking.

3.   Type of IERs

IERs can take several forms. Polystyrene sulfonate is an example of the most common commercial resin [5]. The anion or cation exchange resins are the most common ion exchange resins because of the ion exchanging process. Unlike anion exchange processes, in the cation exchange process, positively charged ions are exchanged like calcium, potassium, sodium, magnesium, and other minerals. With anion exchange resins, ions like sulfate, chloride, fluoride, etc. were exchanged. On the other hand, according to the functional groups present, ion exchange resins are also categorised into four main types such as:- (i) highly acidic ion exchange resins (sodium polystyrene sulfonate on sulphonic acid groups), (ii) highly basic IER (tri methyl ammonium on quaternary amino group),(iii) slightly acidic IER (groups of carboxylic acid) and (iv) slightly basic ion exchange resins (1^0, 2^0, and 3^0 amino groups, like poly ethylene amine). There are two types of SBA resins.

Type (I) SBA resins are used where low levels of silica leakage is an important operating criterion or in warmer climates. Type (II) SBA resins have an exchange site that is chemically weaker than Type (I) resins. Type (II) SBA resins have the advantage of a higher initial exchange capacity. Some resins (iminodiacetic acid and thio-urea-based resins etc.) also form chelate complexes with metal ions to separate metal ions from various mixtures or solutions and such specialised resins are called chelating resins.

Three types of resins such as isoporous, macroporous and gel are also characterised [6]. Macro-porous resins have better stability, physical strength, longer life and resistance to attrition, organic fouling and oxidation. For normal softening and less stringent demands, gel resins should definitely be considered. A gel type resin bead indicates a continuous phase since there is no true porosity, while a macro-porous resin will show a discontinuous phase since each bead is made up of a number of microspheres. A comparative study of the three types of resin is given in table 1 [6].

Table 1. Comparative study of isoporous, macroporous and gel types IERs.

PARAMETERS	GEL	ISOPOROUS	MACROPOROUS
Structure	Non uniform, heterogeneous structure	Uniform, homogeneous Structure	Heterogeneous structure With large cavities
Pore size	Very small ($<40°A$)	Larger than gel	Very large pore size (Avg. $1300°A$)
Resistance to Organic fouling	Gets easily fouled with Organics	Excellent resistance to Organic fouling	Very good resistance To organic fouling
Operating Capacity	Normally higher than macro porous but lower than isoporous type.	Higher than gel and microporous types	Lower than gel and isoporous types
Operating Cost	Medium	Low	High
Regeneration Efficiency	Good	Excellent	Less
Treated Water Quality (Silica leakage)	Average	Excellent	Good
Osmotic Stability	Good	Excellent	Good
Elasticity	Low	High	Very Poor

4.   Synthesis of IERs

Polymerisation and functionalization are two simple steps to synthesize IERs [7,8]. The polymerization step involves two distinct polymer precursors and produces a synthetic functional organic polymer material (cross linked or copolymer). Then the functionalization step introduces an anionic or a cationic part to produce an acid (cation) or a base (anion) exchange site on the resin molecules respectively. Usually the ion-exchanging sites (cation or anion) are introduced after the polymerisation step. These ion exchange sites can be exchanged with metal ions. The synthesis of an anion exchange resin i.e., polymerization and functionalization of an anion exchange resin from chloromethyl styrene copolymerized with divinylbenzene is shown in figure 1 [7].

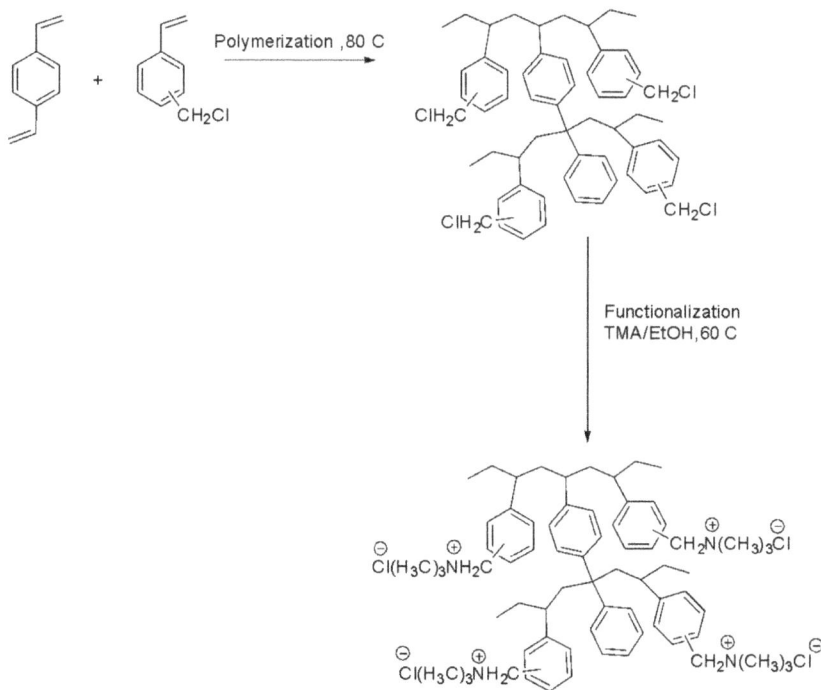

Figure 1. Synthesis of IERs

5.   Uses of IERs

Ion-exchange resins are broadly applied for various works in a large number of fields such as for water purification, water softening, metal separation, juice purification, manufacturing of pharmaceuticals, catalysis, sugar manufacturing and carbon dioxide capture from ambient air etc. [4]. More specifically, ion exchange resin functions as an exchanger, selector, adsorbent, and a catalyser. Nowadays for industries and other purposes, water purification and softening ion exchange resins are very commonly used. It is also quite common to employ ion-exchange resins as substituents of natural and manufactured zeolites. Filtering biodiesel with ion-exchange resins is also highly effective. It is generally used to recover metals, heavy metals, non-metals, purify harmful ions, and remove alkaline or acidic organic substances in effluent and industrial waste water treatment. Because of their high surface area, insolubility in water, as well as appropriateness in steam and fluid processes, ion exchange resins are also utilized as catalysts in certain organic synthesis. Resins are indeed employed in the production of fruit drinks like orange juice and cranberry, in which they are utilized to get rid of bitter components and enhance the flavour [4]. They are also applied to get high valued metals out of minerals, ores, sludge, sea water, including industrial by-products. Floating agents, depressants, flocculants, and collectors are all utilized with them. It's worth noting that these resins are quite useful in situations when it's necessary to concentrate or eliminate materials that are present in extremely low quantities.

6.   Activity of IERs

Resins used for ion exchanges are highly active compounds. Due to the relatively low cost and extensive application ranges, they are commonly utilized for the determination of numerous materials (such as the isolation and prior concentration of radioactive elements from native ores or minerals). The charge, acidity, and basicity of such active groups all improve the activities of ion exchange resins. Strongly acidic sulfonated resins with $-SO_3$ groups are more active than less acidic carboxylated and phosphorylated resins among the most regularly used cation exchangers. With increasing the charge and decreasing hydrated ion radius, the affinity of metal ions or cations for ion exchange sites increases. Anion exchangers usually have amines as functional groups, and strongly basic anion exchangers having substituted quaternary amines are more active than resins with moderately basic 1^0, 2^0, and 3^0 amino groups. Another way, the cross linking reduces the resin's ion-exchange capacity and lengthens the time required to complete ion-exchange procedures while increasing the resin's resilience [9]. Particle size has an impact on resin qualities; finer particles have a greater exterior surface however they create higher heat loss in column processes. Ion exchange resins can also be fashioned into membranes in addition to bead-shaped materials. When electrodialysis is performed, ions pass through ion-exchange membranes, but not water. They're made up of highly crosslinked Ion-exchange resin.

High cross linking resins (12-16 % DVB) are highly expensive to manufacture as well as operate [9]. They were designed with large scale industrial applications in view. These materials are more resistant to oxidizing chemicals like chlorine, as well as physical pressures that can break down lighter-duty materials. The particle size distribution in typical ion exchangers is in the 20-50 mesh range (to distinguish anion and cation/ionic species and nonionic species). The particles must be finer or the degree of crosslinking should be lower for more difficult separations. Furthermore, a grain size of 200-400 mesh is needed when the separation is based solely on minute variations in ion affinity, whereas a molecule size of 50-100 mesh is acceptable when the complexing agents increase selectivity. For analytical purposes, ion exchangers having less than 100 mesh size are utilized, while materials with a mesh size of less than 50 mesh are employed for commercial applications.

7.   Properties of IERs

7.1   IE capacity of resin

The ion exchange process depends on the functional groups of ion exchangers. Calculation of the ion exchange ability of a resin is done based on how many ion exchange sites are present in each unit weight or volume [10]. The amount of substance required to combine with one millimole of hydrogen ions is generally stated in milliequivalent per gram (meq/g) of dry resin or (meq/ml) of hydrated resin, where meq is the amount of substance required to combine with one millimole of hydrogen ions. For a given volume or weight of resin based on the resin capacity, the number of ion exchange sites needed for a particular amount of sample can be calculated. Resins with a variety of cross-linkages generally maintain relatively stable capacity but hydrated resins with weak cross-linkages experience significantly lower capacity. When loading a column with the ion exchanger, keep in mind that the resin volume might fluctuate dramatically depending on the counter ions. If the resin does not have enough room to expand, high pressure can generate. When the resin reaches saturation, the ion exchange behaviour and capacity change; (Bio-Rad b) exhibits an anion exchange capacity of 1.2 meq/mg while cation exchange resins (Bio-Rad a) exhibit an ion exchange capacity of 1.7 meq/mg.

7.2   Water retention capacity of ion exchange resin

It is the amount of water associated with one gram of dry resin. It indicates the extent of cross-linking in the resin. Higher the water retention capacity (WRC), lower the cross-linking, while low water retention capacity shows slow operating capacity, poor elasticity, poor kinetics, brittleness and poor osmotic shock resistance [9].

7.3   Density of ion exchange resin

This is the specific gravity of the resin. Density of resin is important in assessing: maximum concentration of chemicals (HCl, H_2SO_4 or NaOH) that can be used for regeneration, separation of anion and cation resins in mixed bed units and also helps to determine the shipping weight etc.

7.4   Surface area of ion exchange resin

This is an important property of the resin. Gel type and isoprous resins have negligible surface area ($<0.01 m^2/g$). Macroporous resin, on the other hand, has a very large internal surface area (at least 10 m^2/g and at times 30-40 m^2/g). The large surface area of macroporous resin is due to the presence of macro-pores within each bead.

7.5   Regeneration of ion exchange resin

The regeneration is a process where resin is converted to the desired ionic form. After exhaustion, the cation exchanger resin can be regenerated with diluted HCl or H_2SO_4 acid and anion exchanger resin can be regenerated with diluted NaOH solution (5%).

8.   Selectivity of IERs

Because of their selectivity, IERs are widely employed for a variety of purposes. When it comes to metal ions, they have a high selectivity. Selectivity is usually determined by the resin's structure, specifically the ionic radius as well as ionic charge density of such metallic ions [11]. Ions with a greater charge density tend to remove those with a lower one. Adsorption of trivalent ions takes place first, followed by divalent ions, and finally monovalent ions. Bond strength governs selectivity between ions of the similar charge. The electrovalent feature of the metal ion determines this. More electrovalent metal ions bind more firmly and are adsorbed selectively. When an effluent contains copper and zinc ions, a solution containing dissolved copper and zinc will preferentially adsorb cupric ions. An ion exchanging mechanism can be understood by using the equations of mass action, where a selectivity coefficient of the resins may be determined:

$$x\ RA^{y+} + y\ B^{x+} \quad = \quad y\ RB^{x+} + x\ A^{y+}$$

$$K_A{}^B = \frac{(RB)^{xy}(A)^{xy}}{(RA)^{xy}(B)^{xy}}$$

Ion Exchange Resins: Biomedical and Environmental Applications   Materials Research Forum LLC
Materials Research Foundations 137 (2023) 55-74   https://doi.org/10.21741/9781644902219-4

K_AB is the coefficient of rational selectivity that is applied to measure how well one ion performs in comparison to another. It usually shows on a relative scale, with a "base" cation as the starting point. The selectivity scale of certain metal ions is shown in Table 2 [12].

Table 2. RSC of a Sulfonated Polymeric (Styrene and Divinylbenzene) Resin for certain Metal Ions

Cation	RSC	Cation	RSC
Lithium (I)	1.00	Cu^{2+}	3.85
Hydrogen ion (H^+)	1.27	Cd^{2+}	3.88
Sodium (I)	1.98	Ni^{2+}	3.93
Uranium Dioxide ion (UO_2^+)	2.45	Ca^{2+}	5.16
Manganese (I)	2.75	Sr^{2+}	6.51
Potassium (I)	2.90	$Ag+$	8.51
Rubidium (I)	3.16	Hg^{2+}	9.31
Caesium (II)	3.25	Pb^{2+}	9.91
Magnesium (II)	3.29	Ba^{2+}	11.55
Zinc (II)	3.47	Tl^+	12.4
Cobalt (II)	3.74		

As the saturation profile of a resin is obtained by successively exposing it to a freshly prepared solution of metal ions, the selectivity of the resin can be determined. In real-world applications, the resin is gradually saturated, using this approach. When the resin is first wet, it likes to take ions in an indiscriminate manner. However, as the resin becomes saturated, it normally liberates some of the ions that were adsorbed at the beginning. Ion C has the greatest affinity for adsorption by this ion [12].

The distribution coefficient is calculated by dividing the metal loading by its equilibrium concentration. Those two metals' separation factor is determined by the ratio of their distribution coefficients. Net resin charge is not considered when calculating the separation factor, so it is different from selectivity coefficient. A resin's practical application is dependent solely on its separation factor.

9. Toxic metals

Metals are all toxic and injurious when they exceed the set limits, but some metals are called toxic metals when they cause very harmful effects even when present in very low concentrations. Generally, the metals are harmful for humans or animals or plants when they enter the body and the entrance of the metals occurs through foods or drinking water. Cadmium, mercury, selenium, zinc, lead, arsenic, chromium and copper etc. are the common toxic metals of great toxicological concern. The probable sources of heavy metals include waste from mines, agricultural runoff, lead acid batteries, occupational exposure, paints, colours, wastes of chemical laboratories, thermal power plant and other industrial wastes etc. which then contaminate our food and drinking water [12]. Metals can adversely affect several organs in the body in acute and chronic ways. A toxic effect of heavy metals can cause disorders such as nervous system disorders, gastrointestinal and kidney dysfunctions, wrinkles, vascular damage, impaired immunity, birth defects, and cancer and others [1,2,12,13].

In addition to high blood pressure and anaemia, people exposed to lead for an extended period are at a higher risk of developing serious haematological damage, brain damage, and liver, lung, spleen, and kidney malfunctions. As a consequence of lead exposure, the kidney, liver, lungs, and spleen become severely affected. Drinking water should not contain more than 0.05 mg/l of lead. Battery production, lead glass production, pigment production, match production, fuel production, photography, and explosives use lead as raw material, but lead pollution leads to environmental harm from these products. Lead damage to the nervous system is manifested primarily by organic lead compounds rather than inorganic lead compounds. People exposed to cadmium compounds can also suffer from respiratory problems and kidney problems. A skeleton that is pathologically altered, loss of smell, extremities pain, hypochromia, and trouble in walking are the other symptoms that appear. In addition, cadmium exposure also leads to 'Itai-Itai' disease. Drinking water should contain a maximum of 0.003 mg/l of Cadmium. Metals like mercury contribute to poisoning of the body. Mercury, a naturally occurring element, is converted into hazardous divalent compounds (methyl and dimethyl mercury) as a result of chemical reactions and biological processes. Most common 'minamata' disease is caused by mercury pollution. When there is inorganic mercury pollution, it can cause burning of the throat, lung tissue damage, vomiting, necrosis of the intestinal mucosa, bloody diarrhoea, and kidney destruction, which can cause anuria and uraemia. Drinking water is allowed to contain only 0.005 mg/l of it. Chromium generally occurs in the form of Cr (IV) or Cr (III) in natural environments. It is a source of valuable material and a critical micronutrient, but when levels are exceeded, it becomes a dangerous pollutant. Industrial effluents (like wastes from metallurgical processes, leather tanning, paint, textile, dyeing, pigments, and steel production) pollute the environment, whereas natural resources of chromium comprise volcanic exhalations, worn rocks, and other biogeochemical processes, among others. For drinking water, the maximum limit is 0.05 mg/dm^3. Acute poisoning is characterized by vomiting, severe abdominal pain, substantial internal organ damage, bloody diarrhoea, observed gastrointestinal ulcers, persistent problems in the body, and severe kidney destruction with haematuria leading to anuria when too much chromium or its component is consumed. Arsenic as well as its components can cause cancer in the skin and the respiratory system, as well as neoplastic diseases in body organs. Arsenic compounds enter the body system through arsenic contaminated meals and water, where they enter the gastrointestinal tract, skin, and respiratory system. Organic arsenic compounds (such as As_2O_3, AsH_3) are less hazardous than inorganic arsenic compounds (such as As_2O_3, AsH_3). Zinc is a valuable and widely used metal in a variety of sectors. The natural mineral zinc, residential wastes, coal, petroleum, and waste from other industries are all sources of zinc pollution. It aids in the correct functioning of living organisms, protein and carbohydrate metabolism; nevertheless, large quantities of zinc disrupt several biochemical processes, resulting in zinc deposits in the kidneys, liver, and gonads. In comparison to other transition metals, nickel is a relatively hazardous element. Nickel carbonyl compounds are the most poisonous nickel compounds. Allergies, mucous membrane damage, protein metabolism abnormality in plasma, dermatitis, chromosome alterations, harmful respiratory effects, changes in bone marrow, and cancer are all symptoms of too much nickel being breathed. Copper pollution occurs as a poisonous metal in the earth's crust, urban and industrial wastes, and other places. Although copper is a necessary metal for plants and animals, excessive amounts can cause harmful effects such as liver destruction as well as gastrointestinal disorder. "Wilson's disease" and "Indian Childhood Cirrhosis (ICC)" are caused due to the aggregation of copper-compounds inside liver cells. Copper salt intake has been linked to skin inflammation, conjunctivitis, gastrointestinal problems, ulceration as well as corneal opacity, nasal septum, gastritis, diarrhoea, and chronic lung damage, among other things. In drinking water, the maximum limit is 0.05 mg/l. Apart from these metals, there are a plethora of other metals and metal compounds that have already polluted our world to varying degrees; therefore, we must think about these toxic metals and separate them from various wastes from industries, mines, municipalities, and agricultural runoff in order to reduce their adverse effect on human health and also on the environment.

10.   Selective separation of toxic metals

First, a resin's selectivity is determined by the characteristic functional groups it contains, which is the reason why strong bases must be specified (e.g., -N+R$_3$), weak bases (e.g., -

$N+R_2H$, $-N+RH_2$), and acids must be specified [14]. The capacity of strongly acidic cation exchangers, which reach maximum performance over a wider pH range, increases with high pH levels, whereas slightly acidic cation exchangers reach their maximum exchange capacity at pH > 7.0. Actually, different ion exchange resins have different selectivity and affinity for metal ions. Depending on the scale of cations and their charge, sulfonic acid resins have different affinity for cations [14]. The affinity for cations increases as the charge of the cation increases: $Na^+ < Ca^{2+} < Al^{3+} < Th^{4+}$.

Despite an increase in affinity among cations of the same charges with increasing atomic number: Lithium ion (I) < Hydrogen (I) < Sodium (I) < Ammonium (I) < Potassium (I) < Rubidium (I) < Caesium (I) < Silver (I) < Thallium (I)

Magnesium (II) < Calcium (II) < Strontium (II) < Barium (II)

Aluminium (III) < Iron (III)

Large ions with high valences have a higher affinity in general. Affinity series of the highly acidic cation exchanger is as below [15,16,17]:

Plutonium (IV) >> Lanthanum (III) > Caesium (III) > Praseodymium (III) > Neodymium (III) >> Samarium (III) >> Europium (III) >> Gadolinium (III) >> Terbium (III) >> Dysprosium (III) >> Holmium (III)>> Erbium (III) >> Thulium (III) >> Ytterbium (III).

Lutetium (III) > Yttrium (III) > Scandium (III) > Aluminium (III) >> Barium (II) > Lead (II) > Strontium (II) > Calcium (II) > Nickel (II) > Cadmium (II) > Copper (II) > Cobalt (II) > Zinc (II) > Magnesium (II) > Uranium dioxide (II) >> Thallium (I) >> Silver (I) > Caesium (I) > Rubidium (I) > Potassium (I) > Ammonium (I) > Sodium (I) > Hydrogen (I) > Lithium (I).

The affinity for Lewatit SP-112 resin is as follows [18]:

Barium (II) > Lead (II) > Strontium (II) > Calcium (II) > Nickel (II) > Cadmium (II) > Copper (II) > Cobalt (II) > Zinc (II) > Iron (II) > Magnesium (II) > Potassium (I) > ammonium (I) > Sodium (I) > Hydrogen (I).

It is interesting to note that the carboxylic functional groups are in the reverse series of affinity for alkali as well as alkaline earth metal ions: Hydrogen (I) > Magnesium (II) > Calcium (II) > Strontium (II) > Barium (II) > Lithium (I) > Sodium (I) >Potassium (I) > Rubidium (I) > Caesium (I).

Modern ion exchange method is used to recover, isolate, and purify metals including uranium, thorium, gold, silver, copper, platinum, chromium, zinc, cobalt, tungsten, and nickel on a commercial scale (large scale). Ion exchange resin is also utilized for very small-scale separation and purification of some noble metals and rare elements of earth, in which the values of retrieved metals are extremely high.

Because of its low cost and simple processing procedures, the application of IER in hydrometallurgy is becoming increasingly popular. To stimulate selectivity towards alkali metals, alkaline earth metals, and heavy metals, chelating ligands are introduced to eliminate specific metal ions. Chelate rings are usually formed by functional groups including O, N, and S. Nitrogen can be found in a variety of forms, including nitroso, nitro, diazo, azo, amide, nitrile, and various amines.

Carbonyl, phenolic, ether, carboxylic, hydroxyl, phosphoryl, and other groups are common forms of oxygen. Sulphur can be found in thioether, disulphide, thiol, and thiocarbamate groups, among other things. By polymerizing appropriate monomers, these groups of oxygen, nitrogen, and sulphur can be included into the polymer surface of resin [19,20]. The acid base characteristics, polarizability, stability selectivity, kinetic, sorptive capacity, and other physicochemical properties of the resin all influence the selection of an effective chelating resin for selective partition of metal ions. Based on the characteristics of functional groups, pH of the solution and the characteristics of chelating ion exchangers, they display sorption ability. In addition to the spatial configuration and position of functional groups, and their contact steric properties, the matrix's properties determine the selectivity of functional groups for metal ions. Iminodiacetate, amidoxime, dithiocarbamate, bis picolylamine, 8-hydroxyquinoline, diphosphonic, aminophosphonic, sulphonic, carboxylic acid groups, isothiourea thiol, and thiourea, among others, are functional groups of some of the most significant resins for selective separation. These functional groups' ion exchangers have strong selectivity, binding energy, and mechanical stability, as well as a pre-concentration factor, facile recovery for repeated sorption-desorption rotation, as well as better sorption reproducibility. Some chelating ion exchangers can selectively separate harmful metals, as mentioned later on.

A chelating metal ion exchanger with hydroxamic as well as amidoxime functional groups is required to separate different metal ions. Duolite ES-346 contains the malonic acid dihydroximate polymer in combination with styrene-divinylbenzene to remove uranium (VI) from marine water and Arsenic (III) from hydrated solutions. At pH 5.5, Iron (III), Copper (II), as well as Zinc (II) adsorbs the most; at pH 6.0, Tungsten (VI), Uranium (VI), Cobalt (II), and Nickel (II) adsorb the most; and at pH 6.5, Cadmium (II) adsorbs the least. At pH 5.5, it might be used to separate Cobalt (II) and Nickel (II) from Copper (II). According to the affinity series, iminoceticdihydroximate resin is also used to separate Uranium (VI), Iron (II), and Copper (II). The sequence is: Uranium dioxide (II) > Iron (III) > Copper (II) > Zinc (II) > Cobalt (II) > Cadmium (II) > Nickel (II) > Zinc (II). The chelating resins that possess amidoxime functional groups, such as Duolite ES-346 and

Chelate N, can be used to concentrate alkali and alkaline earth metal ions.These resins are especially effective in concentrating solutions having silver (I), Aluminium (III), Cadmium (II), Cobalt (II), Chromium (III), Copper (II), Iron (III), Mercury (II), Manganese (II), Nickel (II), Molybdenum (VI), Lead (I). The selectivity series for amidoxime resins was discovered as: Copper (II) > Iron (III) > Arsenic (III) > Zinc (II) > Nickel (II) > Cadmium (II) > Cobalt (II) > Chromium (III) > Lead (II). It is utilized for initial concentration of Zinc (II) and Lead (II) (III) because salicylic acid has the ability to selectively bind to Zinc (II), Lead (II), and Iron. Salicylaldoxime and salicylaldehyde, on the other hand, are functional chelating groups in phenol-formaldehyde resins that have a higher selectivity for Copper (II) ions. The following are the affinity sequences of salicylaldoxime metal complexes: $Fe^{3+} > Cu^{2+} > Ni^{2+} > Zn^{2+} > Co^{2.}$

According to recent studies, Cs(I) and Sr(I) are particularly attached to IER with phenolic groups in aqueous radioactive solutions. 8-hydroxyquinoline appears to have an impact on both the competitive sorption and separation of intra-groups for phenolic resins with a molecular matrix containing 8-hydroxyquinoline. On the basis of the separation factors obtained by phenolic IER from aqueous solutions, ion-specific resins could be designed to selectively extract actinide ions from nuclear waste.

Ion exchangers containing dithiocarbamate functional groups (Nisso ALM 125), where S is a donor atom, also demonstrate a greater affinity towards transition metal ions. Mercury (II), Lead (II) and Cadmium (II), as well as precious metal ions have a higher affinity for this type of ion exchanger than alkali and alkaline earth metals. Manganese (II), Lead (II), Cadmium (II), Copper (II), Iron (III), and Zinc (II) can also be isolated as well as concentrated using this resin in complicated matrices.

In these experiments, vanadium dioxide, iron (II, III), cobalt (II), nickel (II), and copper (II) were sorbed using poly(iminoethyl)dithiocarbamate copolymers.

Nickel (II), Copper (II), Cobalt (II), Chromium (III), Iron (III), Cadmium (II), Zinc (II), Mercury (II), and Lead (II) are all sorbable using chelating ion exchangers with 8-hydroxyquinoline functional groups. Chelating ion exchangers with iminodiacetic functional groups, which contain two carboxyl groups and a nitrogen atom, have a strong affinity for chromium (III) and copper (II). Chelex 100, Dowex A 1, CR-20, Lewatit TP 207, Purolite S 930, Amberlite IRC 748, or Wofatit MC-50 are some of the industrial chelating ion exchangers and they have the following affinity order: Chromium (III) > Copper (II) > Nickel (II) > Zinc (II) > Cobalt (II) > Cadmium (II) > Iron (II) > Manganese (II) > Calcium (II) >> Sodium (I). Mercury (II) as well as Antimony (V) ions have a high affinity for this sort of ion exchanger. As a result of their iminodiacetate groups, alkaline earth metals can produce stable coordination covalent bonds that are 5000 times stronger than those of alkali metals such as Ca(II). Accordingly, the affinity series for this ion exchanger could be listed as follows:

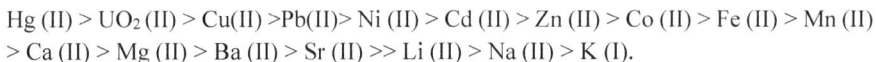

Hg (II) > UO_2 (II) > Cu(II) >Pb(II)> Ni (II) > Cd (II) > Zn (II) > Co (II) > Fe (II) > Mn (II) > Ca (II) > Mg (II) > Ba (II) > Sr (II) >> Li (II) > Na (II) > K (I).

Among the different types of acidic ligand-based chelating ion exchangers, those with phosphoric as well as aminophosphonic functional groups are of specific interest because they have affinity towards heavy metal ions [21]. They also include methylenediphosphonate, ethylenediphosphonate, as well as carboxyethyl phosphonate resins, in addition to phosphate ($-OPO(OH)_2$), phosphinic ($-PO(OH)$), as well as phosphonic ($-PO(OH)_2$) resins. Thorium(IV), Uranium(IV,VI) ions, and Copper(II), Cadmium(II), Zinc(II), Nickel(II), Silver(I), Gold(III), as well as Iron(III) ions, have extraordinarily high selectivity in chelating ion exchangers containing phosphonic functional groups. Diaion CRP200 and DiphonixResin are commercially available resins that contain phosphonic groups.

A selective ion exchange resin contains chelating ion exchange groups composed of the methylglucamine N-methyl (poly hydroxy hexyl) amine functional groups.

Amberlite IRA 743, Dowex BSR 1, Duolite ES-371, Purolite S 108, Diaion CRB 02, and Purolite S110 are a few examples of industrially accessible ion exchangers. These types of ion exchangers can be used to remove Cr(VI) and As(V). These ion exchangers have strong boron selectivity (as Boric Acid H3BO3).

Ion exchange resins containing 2-pyridylmethyl amine functional groups are able to adsorb transition metal ions, especially Cu(II) ions, because of the presence of nitrogen atoms that form connections with Cu(II). Dowex M 4195 is a commercially available product with such functional groups.

Ion exchangers that include thiol, thiourea, and isothiourea groups are essential (Duolite ES-465, Chelate S, Imac GT 73) in addition to thiol groups, they also contain sulfone groups and are nanoporous ion exchangers along PS-DVB matrix. In addition to Hg(II) ions, Silver (I), Gold (III), Platinum (IV), and lead ions are selectively adsorbable using TMR (II).

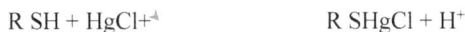

$RSH + Hg^{2+}$ $(R S)2Hg^{2+} + 2H^+$

$R SH + HgCl^+$ $R SHgCl + H^+$

Dowex Retardion 11A8 is a particularly fascinating quaternary ammonium and carboxylic functional group ion exchanger. It is a type of an amphoteric resin. It is possible to separate Cadmium (II) ions from other heavy metals with Dowex Retardion 11A8. Using this resin, an anion exchange takes place in acidic media for the dissociation of Gallium (III), Indium (III), Thallium (III), Platinum (II), Lead (II), from Sodium (I), Nickel (II), Copper (II), Zinc (II) combinations. In solution, Retardion 11A8 can handle cations as well as anions (e.g., Nickel, Cobalt, Copper, and Zinc), it is possible to isolate specific elements more selectively from complicated mixtures than with monofunctional ion exchangers.

Saturated resins, such as Amberlite XAD resins, formed by physically loading organic chemicals in an inert solid support material, are an appealing material for heavy metal ion separation and preconcentration. High porosity, regular pore size, and a large surface area characterize the resins as chemically homogeneous, non-ionic structures.

11.   Modern ion exchange separation method in industry and its future prospects

The use of modern ion-exchange technologies has become essential for many applications, including water softening, environmental remediation, wastewater treatment, hydrometallurgy, chromatography, and biomolecular separations. A considerable number of metals like copper, zinc, chromium, aluminum, as well as lead are seen in industrial waste water, which are largely produced by the electroplating sector. As a result, IER is becoming increasingly important in processes of treating industrial waste water and also in solid acid catalysis processes [22,23]. Similarly, the production of radioactive wastes has increased day by day in tandem with the rapid development of the nuclear sector. So, the treatment of nuclear solid radioactive wastes by IERs is critical, and it has also become a significant problem for the development of current IERs. The IERs are extremely efficient at removing ammonium and phosphorus from synthetic industrial effluents [24,25]. On fruit juice preparing technology, adsorbent resins are very important. They are also excellent for removing colored organic components of pear juice that have been pre concentrated. Color intensity and acidity are greatly decreased by the ion exchanger. This is a great way to keep their nutritional value while also extending their storage life. Food industry applications of ion exchange technology include decolorizing sugar beet and sugar cane extracts during sugar production, increasing the quality of fruit juices and sugar cane syrup, and reducing the bitterness in citrus juices [26]. Protein adsorption is utilized in pharmaceutical purification, immunoassays, biosensors, and therapeutic apheresis. IERs techniques is also very useful in cosmetic manufacture, purification, and surfactant separation [27,28].

Scandium is currently one of the most crucial and demanding metals in modern technologies. But it is very critical because of its great economic relevance, limited availability, as well as low recycling rate. It can be deduced from modern ion exchange techniques that there is a greater orientation in obtaining a highly pure scandium solution by using separation procedures [29,30]. Bio-sorption via ion exchange methods is also significant for the removal as well as recovery of organic inorganic substances, which is important in both environmental and conventional bio-treatment [31]. The current potential impact of ionic pollution on human health and the environment around the world has sparked a lot of interest in ion exchangers for capturing hazardous cationic and anionic species. Moreover, deionization of aqueous solutions, energy conservation and storage, and electrochemical synthesis are also great aspects of ion exchange technique.

Conclusion

A variety of methods for ion exchange have been available for many years. The wide range of resins available today is fuelling interest in ion exchange technology. The right resin selection can make ion exchange a cost-effective and effective pollution control method. Toxic ions in drinking water can be exchanged for other ions via solid ion exchange resin. Nowadays metals are common contaminants in surface water, groundwater, industrial wastewater and other effluent from various sources in the world. Such metals openly challenge and seriously threaten the environment and all living beings. In recent years, huge techniques and instruments are developed to separate and filter the metals from different sources. Although the ion exchange process with resin is rather versatile, it can compete with any other separation method in terms of its applications.

References

[1] J. Briffa, E. Sinagra, R. Blundell, Heavy metal pollution in the environment and their toxicological effects on humans, Heliyon. 6 (2020) e04691. https://doi.org/10.1016/j.heliyon.2020.e04691

[2] M. Balali-Mood, K. Naseri, Z. Tahergorabi, M.R. Khazdair, M. Sadeghi, Toxic Mechanisms of Five Heavy Metals: Mercury, Lead, Chromium, Cadmium, and Arsenic. Front. Pharmacol. 12 (2021) 643972. https://doi.org/10.3389/fphar.2021.643972

[3] ScienceDirect, Ion exchange. https://www.sciencedirect.com/topics/chemical-engineering/ion-exchange (accessed 25 March 2022)

[4] Wikipedia, Ion exchange resin, https://en.wikipedia.org/wiki/Ion-exchange_resin (accessed 25 March 2022)

[5] F. Dardel, T.V. Arden, "Ion Exchangers" in Ullmann's Encyclopedia of Industrial Chemistry, Wiley-VCH, Weinheim. 2008).

[6] Matten Plant, Types of Resins. https://www.mattenplant.com/ion-exchange-ix/overview/types-of-resins (accessed 25 March 2022)

[7] H.A. Ezzeldin, A. Apblett, G.L. Foutch, Synthesis and Properties of Anion Exchangers Derived from Chloromethyl Styrene Codivinylbenzene and Their Use in Water Treatment, International Journal of Polymer Science. 684051 (2010) 9. https://doi.org/10.1155/2010/684051

[8] I.H. Spinner, J. Cirik, W.F.Graydon, Preparation of Ion-Exchange Resins, Canadian Journal of Chemistry. 32 (1953) 143-152. https://doi.org/10.1139/v54-021

[9] The Role of Cross-linking in Ion Exchange Resins (accessed 25 March 2022)

[10] WCP Online, The capacity of ion exchange resin. https://wcponline.com/2017/03/20/capacity-ion-exchange-resin/ (accessed 25 March 2022)

[11] Ion Exchange Resin - an overview | ScienceDirect Topics (accessed 25 March 2022)

[12] S.R. Rao, Resource Recovery and Recycling from Metallurgical Wastes, Pages. 1-581 (2006), ISBN: 978-0-08-045131-2, ISSN: 1478-7482.

[13] J. Briffa, E. Sinagra, R. Blundell, Heavy metal pollution in the environment and their toxicological effects on humans, Heliyon. 6 (2020) e04691. https://doi.org/10.1016/j.heliyon.2020.e04691

[14] A.V. Singh, N. k. Sharma, A.S. Rathore, Synthesis, characterization and applications of a new cation exchanger tamarind sulphonic acid (TSA) resin, Environmental Technology. 33 (4) (2012) 473-480. https://doi.org/10.1080/09593330.2011.579184

[15] M. Draye, K.R. Czerwinski, A. Favre-Reguillon, J. Foos, A. Guy, M. Lemaire, Selective separation of lanthanides with phenolic resins: extraction behavior and thermal stability, Separation Science and Technology. 35 (8) (2000) 1117-1132. https://doi.org/10.1081/SS-100100215

[16] N. Dumont, A. Favre-Reguillon, B. Dunjic, M. Lemaire, Extraction of Cesium from an alkaline leaching solution of spent catalyst using an Ion-Exchange Column, Separation Science and Technology. 31 (7) (1996) 1001-1010. https://doi.org/10.1080/01496399608002501

[17] K.L. Noyes, N. Charton, Micheline Draye1, K.R. Czerwinski, Synthesis and evaluation of resins for actinide separations, Materials Research Society. 757 (2003) 635-640. https://doi.org/10.1557/PROC-757-II11.10

[18] Z. Hubicki, D. Kołodyńska, Selective Removal of heavy metal ions from waters and wastewaters using ion exchange methods, in: A. Kilislioğlu (Eds.), The Edited Volume: Ion Exchange Technologies, IntechOpen, 2012, pp. 193-240. https://doi.org/10.5772/51040

[19] M.A. Ali, M.A. Rahman, A.M. Shafiqul Alam, Use of EDTA-grafted anion-exchange resin for the separation of selective heavy metal ions, Analytical Chemistry Letters. 3 (3) (2013) 199-207. https://doi.org/10.1080/22297928.2013.838454

[20] G. Al-Enezi, M.F. Hamoda, N. Fawzi, Ion Exchange Extraction of Heavy Metals from Wastewater Sludges, Journal of Environmental Science and Health. A. 39 (2004) 455-464. https://doi.org/10.1081/ESE-120027536

[21] R. Kiefer, W.H. Holl, Sorption of heavy metals onto selective Ion-Exchange Resins with aminophosphonate functional groups, Ind. Eng. Chem. Res. 40 (2001) 4570-4576. https://doi.org/10.1021/ie0101821

[22] M.A. Harmer, Q. Sun, Solid acid catalysis using ion-exchange resins, Appl. Catal. Gen. 221 (2001) 45-62. https://doi.org/10.1016/S0926-860X(01)00794-3

[23] S.A. Cavaco, S. Fernandes, M.M. Quina, L.M. Ferreira, Removal of chromium from electroplating industry effluents by ion exchange resins, J. Hazard. Mater. 144 (2007) 634-638. https://doi.org/10.1016/j.jhazmat.2007.01.087

[24] J. Wang, Z. Wan, Treatment and disposal of spent radioactive ion-exchange resins produced in the nuclear industry, Prog. Nucl. Energy. 78 (2015) 47-55. https://doi.org/10.1016/j.pnucene.2014.08.003

[25] J. Paul Chen, M.L. Chua, B. Zhang, Effects of competitive ions, humic acid, and pH on removal of ammonium and phosphorus from the synthetic industrial effluent by ion exchange resins, Waste Manag. 22 (2002) 711-719. https://doi.org/10.1016/S0956-053X(02)00051-X

[26] J. Kammerer, R. Carle, D.R. Kammerer, Adsorption and Ion Exchange: Basic principles and their application in food processing, J. Agric. Food Chem. 59 (2011) 22-42. https://doi.org/10.1021/jf1032203

[27] L. Kisley, J. Chen, A. P. Mansur, B. Shuang, K. Kourentzi, M. Poongavanam, W. Chen, S. Dhamane, R. Willson, C. F. Landes, Unified superresolution experiments and stochastic theory provide mechanistic insight into protein ion-exchange adsorptive

separations, App. Phy. Sci. 111 (6) (2014) 2075-2080.
https://doi.org/10.1073/pnas.1318405111

[28] R. F. Schubert and P. H. Ko, Methods used in the analysis of shampoos, J. Soc. Cosmet. Chem. 23 (1972) 887-898.

[29] V. R. Moreira, Y. A. R. Lebron, A. F. S. Foureaux, L. V. de S. Santos, and M. C. S. Amaral, Acid and metal reclamation from mining effluents: Current practices and future perspectives towards sustainability, J. Environ. Chem. Eng. 9 (2021) 105169. https://doi.org/10.1016/j.jece.2021.105169

[30] A.B. Junior, D.C.R. Espinosa, J. Vaughan, J.A.S. Tenório, Recovery of scandium from various sources: A critical review of the state of the art and future prospects, Miner. Eng. 172 (2021) 107148. https://doi.org/10.1016/j.mineng.2021.107148

[31] M. Fomina, G.M. Gadd, Biosorption: current perspectives on concept, definition and application, Bioresour. Technol. 160 (2014) 3-14.
https://doi.org/10.1016/j.biortech.2013.12.102

Ion Exchange Resins: Biomedical and Environmental Applications   Materials Research Forum LLC
Materials Research Foundations 137 (2023) 75-92   https://doi.org/10.21741/9781644902219-5

Chapter 5

Separation and Purification of Bioactive Molecules by Ion Exchange

Rabiul Alam[1] and Bidyut Saha[2*]

[1] Department of Chemistry, Rabindra Mahavidyalaya, Champadanga, Hooghly 712401, West Bengal, India

[2] Department of Chemistry, The University of Burdwan, Burdwan 713104, West Bengal, India

* bsaha@chem.buruniv.ac.in

Abstract

Bioactive molecules (or signaling molecules) are involved in tissue regeneration and have an important role in modulating the microenvironment in vivo. Bioactive chemicals have the ability to govern host cell motility, proliferation, and differentiation, as well as enabling cells to interact with their surrounding microenvironment via particular receptors for chemical recognition. This chapter introduces ion-exchange chromatography (IEC) as a method for separating and purifying bioactive compounds such as polyphenols, catechin derivatives from complicated plant mixtures, proteins, minor whey protein, peptides, human C-peptide, alkaloids from Chinese medicines, plasmid DNA and carbohydrates.

Keywords

Ion-Exchange Chromatography, Biochemical Purification, Bioactive Molecules, Polyphenols, Catechin, Proteins, Peptides, Human C-Peptide, Alkaloids, Plasmid DNA, Carbohydrates

Contents

Separation and Purification of Bioactive Molecules by Ion Exchange.........75

1.   Introduction..76

　　　1.1   Reversed phase chromatography 77

2.   Polymeric sorbents for preparative chromatography of biologically active compounds 78

2.1   Designing a biochemical purification 79

3.   **Ion-exchange separation and purification of polyphenols** 79

3.1   Separation of bioactive catechin derivatives by AEC 82

4.   **Ion-exchange separation and purification of protein** 82

5.   **Use of ion-exchange chromatography for the separation of peptide** .. 83

5.1   Separation of human C-peptide by ion exchange 85

6.   **Separation of Alkaloids from Chinese Medicines by ion-exchange** 85

7.   **Separation of plasmid DNA using ion-exchange chromatography** 86

8.   **Separation of carbohydrates from seaweed using ion-exchange chromatography** 86

9.   **Future Prospects** 87

References 88

1.   Introduction

Separation of various substances is achieved by using ion exchange chromatography. Ion exchange chromatography is a technique for separating mixtures of charged compounds such as cations, anions, amino acids, proteins or neutral molecules that can develop a charge in acidic or basic media such as carboxylic acids and amines. Ion exchange chromatographic separation is done by using porous resin beads(granules) to which are bonded acidic groups such as $-SO_3^-H^+$ or $-COO^-H^+$ or basic groups such as $-CH_2N^+R_2X^-$ or $-CH_2NR_2$. Sorting of cations is done employing cation exchange resins, whereas sorting of anions is done with anion exchange resins.

After settling the column with resin, a buffer is pushed through the column to equilibrate the charge. Solute molecules and buffer ions will interchange after the sample is injected. The species (ions or molecules) with low charges elute first, followed by those with increasingly higher charges. The factors such as pH, buffer and temperature, play essential roles in determining the separation.

Fig.1 shows a *cation exchange column*, which is a form of ion exchange column. The support in this example is made up of small beads to which charged chemicals are bonded. A counter-ion exists for each charged molecule which is Na^+ in this case.

Because of their covalent attachment, the negatively charged groups are unable to exit the beads, but the Na^+ ions can be "*exchanged*" with the species having similar charge. Positively charged chemicals in a mixture pass through a cation exchange column and "stick" to the acidic groups on the granules. Neutral or negatively charged molecules in the sample will travel swiftly through the column.

Negatively charged species (molecules or ions) in the sample will adhere, whereas other species will pass across fast. To dislodge and release the species that have become stuck in a column, simply injection of a large concentration of the suitable counter-ions is needed. This approach enables the recovery of all mixture components with similar charges. Ottens and coworkers [1] stressed that an IEC procedure is required for the preparative purification of biomolecules, and that the resins and adsorbents used must be chosen carefully. The major stages in the IEC process are the adhesion and liberation of species of interest. The essential processes in the IEC process are resin equilibration, loading of samples, washing and elution, which can be done in isocratic or gradient mode. Cyclic IEC methods, which require less volume of solvents (30% EtOH) with high recovery efficiency for EGCG was reported [2]. Biomolecules such as anthocyanins [3], polyphenols [4] from strawberries and red wine were purified by the IEC method. Furthermore, using a non-solvent IEC technique may be a preferable solution.

1.1   Reversed phase chromatography

In the purification of biomolecules, reversed phase chromatography has shown to be a crucial technology. Small chemical molecules were separated using this approach, which was developed in the 1960s. The phrase reversed-phase chromatography is employed because the polarity of the mobile and stationary phases has been reversed. To purify biomolecules such as peptides, proteins, and oligonucleotides RPC has been used. RPC is frequently used in conjunction with other chromatographic methods like gel filtration and ion exchange chromatography. The extra techniques that are used, as well as the order in which they are used, are important for successful purification. In research labs, reversed phase chromatography has shown to be quite effective in purifying biomolecules.

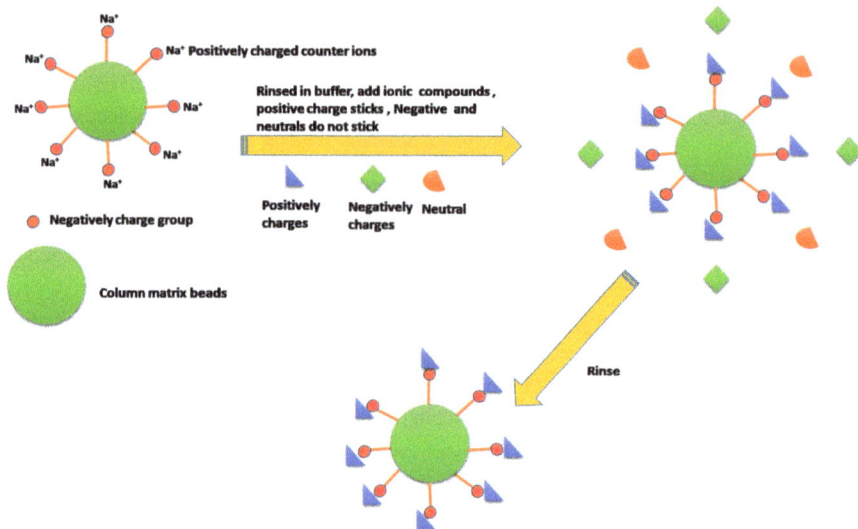

Figure 1. Schematic diagram of a cation exchange column.

2.   Polymeric sorbents for preparative chromatography of biologically active compounds

Currently, a large number of organic polymers are used for the design of chromatographic media. The effective sorbents should possess the following set of properties:

1. High adsorption capacity and fast kinetics of the desired product.

2. Relatively rigid structure, high flow ability, hydrodynamic stability against hydrolysis and mechanical stress.

3. Complete reversibility of selective adsorption of target components in relatively "soft" physico- chemical conditions, which permits the formation of eluates with high concentrations.

4. Absence of functional groups in a sorbent (which lead to both irreversible interactions between sorbent and target BAC and non-specific interactions with impurities).

5. Easy regeneration and multiple reproducibility of chromatographic processes.

The above properties can be achieved with the knowledge of basic trends controlling formation of polymer networks

2.1   Designing a biochemical purification

Sample dilution and impurity contamination are normally at their highest levels at the start of a purification protocol. High-capacity, low-resolution procedures that result in sample concentration are commonly used at this stage. Higher and higher resolution approaches are used at later phases of the process. The overall approach used is influenced by the type of biomolecule to be purified as well as its source. The biomolecules that we have so far concentrated on in this chapter can be classified (Scheme 1) as follows in terms of ion exchange chromatography:

- Ion-Exchange separation and purification of polyphenols.
- Separation of beneficial polyphenol compounds from plant mixtures by IEC.
- Separation of protein by Ion-Exchange.
- Ion-Exchange Separation and Purification of Minor Whey Protein.
- Peptide separation using IEC.
- Human C-peptide separation using IEC.
- Separation of Alkaloids from Chinese Medicines by Ion-Exchange.
- Separation of Plasmid DNA using ion-exchange chromatography.
- Separation of Carbohydrates from seaweed using ion-exchange chromatography

All of these categories differ in a number of aspects that are crucial to their purification. On the following pages, you'll find examples of how ion exchange chromatography has been used to solve some common purification difficulties, as well as some general guidelines for developing successful purification techniques.

3.   Ion-exchange separation and purification of polyphenols

After solvent extraction, polyphenols from red wine were purified using a cation exchange chromatography approach [4]. The use of various solvents as an eluent in IEC is highly required for additional purification of the substance. Dextran based weakly acidic cation exchange gels are used for this purpose [5].

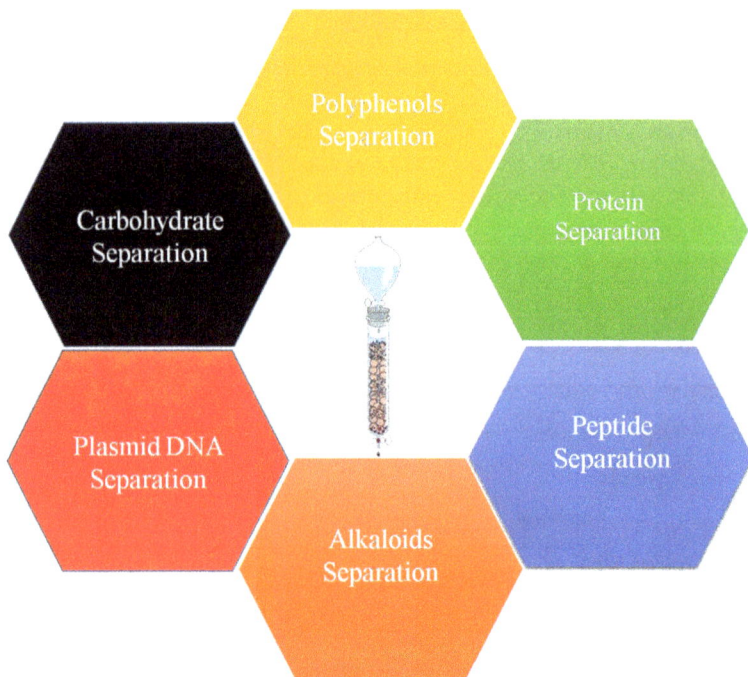

Scheme 1. Separation of various bioactive molecules by ion exchange.

Phenolic chemicals have beneficial effects on human health. So, polyphenol purification technologies are in high demand these days. A lot of study has gone into analyzing biological activity and identifying phenolic compounds, but the extraction procedure is frequently overlooked. Polyphenol extraction and isolation are difficult owing to their degradation and reactivity [6-9]. Phenolic compounds can be found mostly in cell walls, leaves and gymnosperms [10].

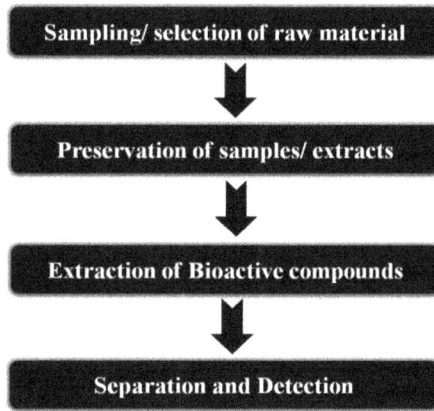

Figure 2. Critical steps for extraction and separation of bioactive molecules.

Furthermore, phenolic groups are covalently linked to plant material, which makes it difficult to liberate them into an extractable form [11]. As a result, several extraction techniques have been developed.

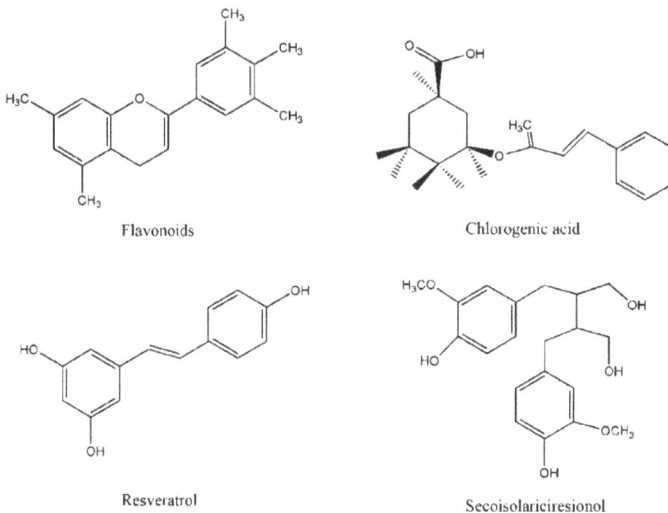

Figure 3. Chemical structure of principal polyphenols.

3.1 Separation of bioactive catechin derivatives by AEC

Anion-exchange chromatography (AEC) was used by Josep L. Torres and colleagues [12] to separate bioactive catechin conjugates. Organic solvents were required to separate the bioactive conjugates from the anion-exchange resin, which was sufficient to remove the poor aqueous solubility contacts with the grid. The NR_n^+(n=4, R=alkyl or aryl) group of the resin interacts with the electron-containing sites of the phenyl ring of the polyphenols via cation–π connections. Chemical structures of catechin and associated conjugates are shown in Fig. 4.

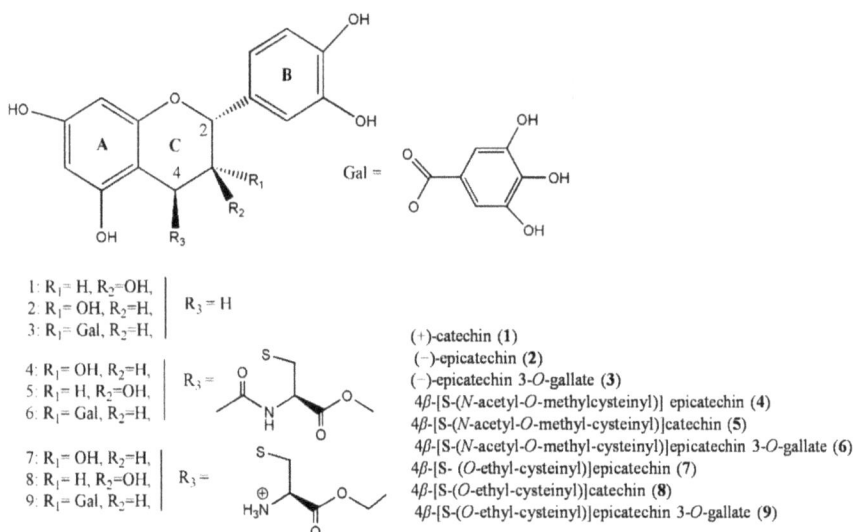

Figure 4. Chemical structures of catechin and associated conjugates.

4. Ion-exchange separation and purification of protein

There is no one-size-fits-all method for purifying all types of proteins [13, 14]. The level of purity necessary is determined by the protein's intended use. Proteins are ampholytes with both positive and negative charges making them complicated ampholytes. The amount of ionizable amino acid residues in a protein's structure determines its isoelectric point (pI), which is the pH at which the net charge is zero. Arginines, lysines, and histidines are commonly used to provide positive charges, depending on the pH of the surrounding buffer. Below pH 8, any free N-terminal amine with a positive charge will likewise contribute.

$C_4H_6NO_4^-$ (aspartate) and $C_5H_8NO_4^-$ (glutamate) residues, as well as the -COOH) group at the C-terminus give the majority of negative charges. All of these residues are ionized above pH=6. Cysteine may also become ionized at higher pH values (>8). Charged groups are almost invariably found on the surface of proteins. Metalloproteins are an exception, where metal is coordinated with charged residues. The cumulative impact of all the charged side chains on the protein will vary depending on the pH of the solute.

As a result, proteins can be separated using an anion exchanger or a cation exchanger with fixed positive or negative charges on the stationary phase. Anion exchange chromatography is performed at pH value beyond the protein's isoelectric point, whereas cation exchange chromatography is performed at pH levels below the isoelectric point. IEC can be performed at a suitable pH range in which the proteins are stable. pH is selected in such a way that it should be at least 1 unit above or below the isoelectric point of the analytes to be separated to obtain effective adsorption. The pH of an ion exchanger's microenvironment differs somewhat from the administered buffer. In anion exchangers, the pH adjacent to the grid is one unit higher than that of the adjacent buffer, while in cation exchangers, it is up to 1 unit lower. As a result, a protein that is bound on a cation exchanger at low pH (5) would be subjected to pH 4, where the protein could be destroyed, if its stability is low. At physiological pH range (pH 6–8), most proteins are negatively charged, an anion exchanger is used as a first step in many IEXC applications. However, for effective separation, the pH should be chosen to introduce as high a charge difference between the protein of interest and the impurities as possible.

As the change in the pH modifies the charge properties of the sample components, IEXC can be used multiple times in a purification strategy. IEXC is typically used to attach the molecule of interest before washing any unbound impurities. If necessary, the procedure can also be used to bind contaminants.

Binding of proteins to ion-exchange matrices is based on the protein's surface charge distribution and ionic strength. Even if the overall charge of the protein is zero, binding can occur, if there are surface areas with large concentrations of charged groups. Furthermore, because protein conformation affects chromatographic behavior, some structural alterations can impact IEXC separation.

5.   Use of ion-exchange chromatography for the separation of peptide

Peptide separation using ion-exchange chromatography is based on specific covalent or non-covalent interactions between the substances and the adsorbate [15]. As net charge of peptides varies with varying the pH of the mobile phase, this approach is applied to peptide purification. Basic protein residues (Arg, Hys, Lys) and $-COO^-$ groups become mostly protonated when the pH of the solutions becomes lower. As a result, IEX is frequently utilized to characterize charge variations of peptides and proteins, despite the fact that, due to its incompatibility with MS detection, it is increasingly being supplanted by Reversed Phase Liquid Chromatography [16].

Ion-exchange chromatography proves to be a suitable option for detecting the modified peptides such as acetylation that are difficult to detect with Reversed Phase Liquid Chromatography. Furthermore, this method can discriminate between analytes of similar hydrophobicity.

Crimmins [17] confirmed that the sequence of elution corresponded to the charge order and during the ion exchange differentiation of an analyte made of synthetic peptides containing +1 - +7 charge at pH value 3; most retained component was the peptide with charge +7 (Fig. 5). The stationary phase was sulfoethyl aspartamide, while the mobile phases were 5 mM Na_3PO_4, 25% CH_3CN and 500 mM NaCl at pH = 3. Different research groups have used the same stationary phase to evaluate peptides (till fifty in a single study, containing between five to twenty amino acid residues) [18-20]. The elution order, on the other hand, does not always correlate to the charge order.

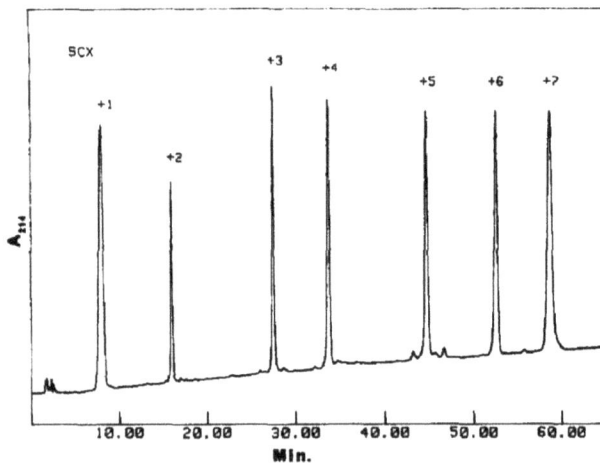

Figure 5. Separation of synthetic peptides with different charges by IEC. Reprinted with permission from ref. [17].

Ion Exchange Resins: Biomedical and Environmental Applications   Materials Research Forum LLC
Materials Research Foundations 137 (2023) 75-92   https://doi.org/10.21741/9781644902219-5

The peptide components which were produced by transformation of myoglobin were studied and an inversion of the elution order was found in relation with the charge [21] proving the peptide's hydrophobicity and at the same time preventing the charged residue due to steric reasons. Pre-purification of the targeted peptide has also been accomplished using ion-exchange chromatography. The purification of C-peptide, for example, was accomplished using the method used for the separation of Lactoferrin on a stationary phase with ammonia solution as an eluent [22]. Because C-peptide has thirty-one amino acid residues, only one of which is basic and has an extremely low isoelectric point of roughly three [23]. For cytochrome-C and hemoglobin digest [16], ion exchange chromatography is regularly used for peptide mapping to illustrate protein recognition.

5.1   Separation of human C-peptide by ion exchange

In patients with diabetes mellitus, C-peptide is frequently used as an indication of insulin secretion levels as C-peptide is synthesized in equimolar concentration with insulin and undergoes less hepatic metabolism. In insulin-treated diabetes, C-peptide measurements can also be used to monitor β-cell activity [24-28]. For C-peptide measurement, some studies have employed the IDA method [29-34]. Unfavorable "matrix effect" has been recognised as among the key reasons of ionization issues and low fragmentation efficiency [32, 33]. To overcome this challenge, Rogatsky and associates presented their 2-D chromatographic technique [33]. A powerful cation exchanger was utilized for the separation of substances from ballast proteins and peptides during C-peptide isolation stage [35]. Theoretical modeling of the gross external charge electricity of the peptide as a consequence of pH was used to optimize the parameters of ion-exchange chromatographic separations. LC–MS/MS was used to further purify and concentrate the material. When compared to the solid matrix separation methods already in use, ion exchange chromatography allowed for the processing of greater sample volumes, which was significant for people with lower levels of C peptide.

6.   Separation of Alkaloids from Chinese Medicines by ion-exchange

Alkaloids are bioactive compounds with a wide range of applications in medicines, cosmetics, and foods. Alkaloids are commonly separated via column chromatography. Ion exchange resins are used to purify alkaloids in the same way as macro porous resins are used, which includes treatment, sampling, washing and elution. Table 1 lists the sample loading, cleaning, elution process, purity, and recovery for various research projects. After sample loading the resin is washed with water. The sample flow rate, pH, solvent, eluent composition and volume are also listed in Table 1.

Table 1. Procedure for the separation of alkaloids by IEC.

Chinese medicines	Alkaloids	Type of Resin	Loading procedure		Cleaning procedure		Elution procedure		Purity (%)	Recovery (%)	Ref
			pH of the sample solution	Sample flow rate	Solvent	Volume of solvent	Eluent composition	Volume of Eluent			
Cynoglossum amabile	Total alkaloids	001x7	Two	Six bed volume /h	H_2O	Five bed volume	$1 mol L^{-1}$ Nacl solution	Twenty six bed volume			38
Corydalis hendersonii		001x7			H_2O	Six bed volume	70% EtOH - 5% NH_4OH	Six bed volume	Greater than forty-five		39
Motherwort		001x7	Two		H_2O		5% NH_4OH 70% EtOH 5% NH_4OH				37
Uncaria rhynchophylla		001x7		Six bed volume /h	H_2O		5% NaCl solution - 50% EtOH	Ten bed volume	~forty-eight	Eighty nine point nine	36
Sini powder		001x7	One				4% NH_4OH-EtOH	Forty mL	zero point three seven three		40

7. Separation of plasmid DNA using ion-exchange chromatography

At pH values above 4, phosphate groups of nucleic acids have a negative charge, which may cause interactions with the positively charged anion exchanger groups. The strength of this interaction depends on the density and shape of the negatively charged groups on the plasmid DNA. Alkaline cell lysate also contains host cell proteins, genomic DNA, and RNA, in addition to plasmid DNA. Because some of these biomolecules are negatively charged and have similar physio-chemical properties like plasmid DNA, they contest for adhesion and are released together with plasmid DNA during AEC. The structure of DNA and the ligand is critical for the binding of large biomolecules like plasmid DNA to ion-exchange ligands. For stronger binding, ligands on flexible arms adopt favorable conformations. Grafting polymer layers onto chromatographic matrices allows for the introduction of specialized qualities generated from the grafted layer while keeping the chromatographic matrix's properties.

8. Separation of carbohydrates from seaweed using ion-exchange chromatography

Among multidimensional methodologies, IEC is a definitive method. IEC separates carbohydrates by adsorbing charged species on the column's solid surface and desorbing them by changing the pH or concentration of the mobile phase [42]. Anion-exchange chromatography is routinely used to isolate carbohydrates from seaweed. AEC is a widely used purification process for separating polysaccharides like fucoidan [42] and carrageenan [43]. Strong anionic charges of the sulfate ester groups attached to these seaweed polysaccharides promote the use of AEC under various experimental settings detailed in Table 2. To create diverse polysaccharide fractions, the adsorbed carbohydrates were eluted from different resins using a stepwise [44] or linear gradient [45-47] of sodium chloride or sodium hydroxide. Furthermore, ion exchange chromatographic separation of fucoidan extracts was used for the determination of the monosaccharide composition of various fucoidan fractions [46].

Table 2. Separation procedure of various carbohydrates.

Compounds	Diagnostic techniques	Experimental settings	Ref.
Alginate	Ion Exchange Chromatography	Immobile part- CarboPac PA1. Eluting liquid-0.1 M, 0.16 M sodium hydroxide and CH₃COONa.	47
Carrageenan IEC		Immobile part- Q. Eluting liquid - 3M NaCl.	46
Fucoidan		Immobile part- DEAE-cellulose. Eluting liquid - 0, 0.2, 0.4, 0.8 and 1.6 M sodium chloride and then 0.3 M sodium hydroxide	44
Laminarin and fucoidan		Immobile part- DEAE-cellulose. Eluting liquid - water , 0–2 M sodium chloride	45

9. Future Prospects

Numerous bioactive compounds have been separated with great success and this movement has been fueled by significant advances in scientific knowledge about their physiological, biochemical and prospective health advantages. Furthermore, a number of uses for these components have been discovered in the food and pharmaceutical sectors. However, large-scale manufacture and use of these components pose a barrier for a variety of reasons. There are also no commercially or economically viable techniques for isolating, fractionating, and concentrating them without compromising their activities or bioavailability.

It is envisaged that the recently established technologies, as well as those under research, would provide better ways for broader and more efficient bioactive compounds such as polyphenols, catechin derivatives from complicated plant mixtures, proteins, minor Whey protein, peptides, human C-peptide, alkaloids from chinese medicines, plasmid DNA and carbohydrates.

References

[1] M. Ottens, S. Chilamkurthi, S. Rizvi, Advances in process chromatography and applications in the food, beverage and nutraceutical industries, in: S.S.H. Rizvi (Eds.), Separation, Extraction and Concentration Processes in the Food, Beverage and Nutraceutical Industries, Woodhead Publishing, Cambridge, UK, 2010, pp. 109-137. https://doi.org/10.1533/9780857090751.1.109

[2] L. Wang, L. -H. Gong, C. -J. Chen, H. -B. Han, H. -H. Li, Column- chromatographic extraction and separation of polyphenols, caffeine and theanine from green tea, Food Chem. 131 (2012) 1539-1545. https://doi.org/10.1016/j.foodchem.2011.09.129

[3] O.M. Andersen, T. Fossen, K. Torskangerpoll, A. Fossen, U. Hauge, Anthocyanin from strawberry (Fragaria ananassa) with the novel aglycone, 5-carboxy pyrano pelargonidin, Phytochemistry. 65 (2004) 405-410. https://doi.org/10.1016/j.phytochem.2003.10.014

[4] X. Vitrac, C. Castagnino, P. Waffo-Téguo, J. -C. Delaunay, J. Vercauteren, J. -P. Monti, G. Deffieux, J. -M. Mérillon, Polyphenols newly extracted in red wine from Southwestern France by Centrifugal Partition Chromatography, Journal of Agricultural and Food Chemistry. 49 (2001) 5934-5938. https://doi.org/10.1021/jf010522d

[5] L. Feng, F. Zhao, Separation of polyphenols in tea on weakly acidic cation-exchange gels, Chromatographia. 71 (2010) 775-782. https://doi.org/10.1365/s10337-010-1545-6

[6] Vr. Cheynier, Polyphenols in foods are more complex than often thought, Am. J. Clin. Nutr. 81 (2005) 223S-229S. https://doi.org/10.1093/ajcn/81.1.223S

[7] R. Tsao, Chemistry and biochemistry of dietary polyphenols, Nutrients. 2 (2010) 1231-1246. https://doi.org/10.3390/nu2121231

[8] M. Naczk, F. Shahidi, Extraction and analysis of phenolics in food, J. Chromatogr. 1054 (2004) 95-111. https://doi.org/10.1016/S0021-9673(04)01409-8

[9] A. Mustafa, C. Turner, Pressurized liquid extraction as a green approach in food and herbal plants extraction: A review, Anal. Chim. Acta 703 (2011) 8-18. https://doi.org/10.1016/j.aca.2011.07.018

[10] P. Hutzler, R. Fischbach, W. Heller, T.P. Jungblut, S. Reuber, R. Schmitz, M. Veit, G. Weissenböck, J. -P. Schnitzler, Tissue localization of phenolic compounds in plants by confocal laser scanning microscopy, J. Exp. Bot. 49 (1998) 953-965. https://doi.org/10.1093/jxb/49.323.953

[11] G. Xu, X. Ye, J. Chen, D. Liu, Effect of heat treatment on the phenolic compounds and antioxidant capacity of citrus peel extract, J. Agric. Food Chem. 55 (2006) 330-335. https://doi.org/10.1021/jf0625171

[12] C. Lozano, J. Bujons, J. C. Josep Llu'ıs Torres Janson, L. Ryden, Novel separation of bioactive catechin derivatives from complex plant mixtures by anion-exchange chromatography, Separation and Purification Technology. 62 (2008) 317-322. https://doi.org/10.1016/j.seppur.2008.01.032

[13] J.C. Janson, L. Ryden, Protein purification, principles, high-resolution methods, and applications, Second ed., Wiley-Liss, New York (1998).

[14] A. Jungbauer, C. Machold, R. Hahn, Hydrophobic interaction chromatography of proteins. III. Unfolding of proteins upon adsorption, J. Chromatogr. A 1079 (2005) 221-228. https://doi.org/10.1016/j.chroma.2005.04.002

[15] M. I. Shaik, N. M. Sarbon, A review on purification and characterization of anti-proliferative peptides derived from fish protein hydrolysate, Food Rev. Int. 38 (2020) 1389-1409. https://doi.org/10.1080/87559129.2020.1812634

[16] S. Fekete, A. Beck, J. L. Veuthey, D. Guillarme, Ion-exchange chromatography for the characterization of biopharmaceuticals, J. Pharm. Biomed Anal. 113 (2015) 43-55. https://doi.org/10.1016/j.jpba.2015.02.037

[17] D. L. Crimmins, Strong cation-exchange high-performance liquid chromatography as a versatile tool for the characterization and purification of peptides. Anal. Chim. Acta. 352 (1997) 21-30. https://doi.org/10.1016/S0003-2670(97)00091-3

[18] A. J. Alpert, Hydrophilic-interaction chromatography for the separation of peptides, nucleic acids and other polar compounds, J. Chromatogr. A 499 (1990) 177-196. https://doi.org/10.1016/S0021-9673(00)96972-3

[19] A. J. Alpert, P.C. Andrews, Cation-exchange chromatography of peptides on poly(2-sulfoethyl aspartamide)-silica, J. Chromatogr. A 443 (1988) 85-96. https://doi.org/10.1016/S0021-9673(00)94785-X

[20] D. L. Crimmins, J. Gorka, R. S.Thoma, B. D. Schwartz, Peptide characterization with a sulfoethyl aspartamide column, J. Chromatogr. A 443 (1988) 63-71. https://doi.org/10.1016/S0021-9673(00)94783-6

[21] D. L. Crimmins, R.S. Thoma, D.W. McCourt, B.D. Schwartz, Strong-cation-exchange sulfoethyl aspartamide chromatography for peptide mapping of Staphylococcus aureus V8 protein digests, Anal. Biochem. 176 (1989) 255-260. https://doi.org/10.1016/0003-2697(89)90305-9

[22] I. Recio, S.Visser, Two ion-exchange chromatographic methods for the isolation of antibacterial peptides from lactoferrin: In situ enzymatic hydrolysis on an ion-exchange membrane, J. Chromatogr. A 831 (1999), 191-201. https://doi.org/10.1016/S0021-9673(98)00950-9

[23] A.V. Stoyanov, C. L. Rohlfing, S. Connolly, M. L. Roberts, C. L. Nauser, R.R. Little, Use of cation exchange chromatography for human C-peptide isotope dilution-Mass spectrometric assay, J. Chromatogr. A 1218 (2011) 9244-9249. https://doi.org/10.1016/j.chroma.2011.10.080

[24] H. Kuzuya, P.M. Blix, D.L. Horwitz, A.H. Rubinstein, D.F. Steiner, C. Binder, O. K. Faber, Heterogeneity of circulating human C-peptide, Diabetes. 27 (1978) 184. https://doi.org/10.2337/diab.27.1.S184

[25] D. F. Steiner, Proinsulin and the biosynthesis of Insulin, N. Engl. J. Med. 280 https://doi.org/10.1056/NEJM196905152802008

1 (1969) 1106-1113.

2 [26] A. H. Rubinstein, J. L. Clark, F. Melani, D. F. Steiner, Secretion of Proinsulin

C-peptide by pancreatic β Cells and its circulation in blood, Nature. 224 (1969) https://doi.org/10.1038/224697a0

697-699.

[27] K. S. Polonsky, J. L. -Paixao, B. D. Given, W. Pugh, P Rue, J Galloway, TKarrison, B Frank, . Use of biosynthetic human C-peptide in the measurement of insulin secretion rates in normal volunteers and type I diabetic patients, J. Clin. Invest. 77 (1986) 98-105. https://doi.org/10.1172/JCI112308

[28] K. S. Polonsky, B. D. Given, L. Hirsch, E. T. Shapiro, H. Tillil, C. Beebe, J. A. Galloway, B. H. Frank, T. Karrison, E. Van Cauter, Quantitative study of insulin secretion and clearance in normal and obese subjects, J. Clin. Invest. 81 (1988) 435-441. https://doi.org/10.1172/JCI113338

[29] A.D. Kippen, F. Cerini, L. Vadas, R. Stöcklin, L. Vu, R. E. Offord, K. Rose,Development of an isotope dilution assay for precise determination of insulin, C-peptide, and proinsulin levels in non-diabetic and type II diabetic individuals with comparison to immunoassay, J. Biol. Chem. 272 (1997) 12513-12522. https://doi.org/10.1074/jbc.272.19.12513

[30] C. Fierens, L. M. Thienpont, D. Stockl, E. Willekens, A. De Leenheer, Quantitative analysis of urinary C-peptide by liquid chromatography-tandem mass spectrometry

with a stable isotopically labeled internal standard, J. Chromatogr. A. 896 (2000) 275-278. https://doi.org/10.1016/S0021-9673(00)00717-2

[31] C. Fierens, D. Stockl, D. Baetens, A.P. De Leenheer, L.M. Thienpont, Application of a C-peptide electrospray ionization-isotope dilution-liquid chromatography-tandem mass spectrometry measurement procedure for the evaluation of five C-peptide immunoassays for urine, J. Chromatogr. B Analyt. Technol. Biomed. Life Sci. 792 (2003) 249-259. https://doi.org/10.1016/S1570-0232(03)00268-X

[32] C. Fierens, D. Stockl, D. Baetens, A.P. De Leenheer, L.M. Thienpont, Standardization of C-peptide measurements in urine by method comparison with isotope-dilution mass spectrometry, Clin. Chem. 49 (2003) 992-994. https://doi.org/10.1373/49.6.992

[33] E. Rogatsky, B. Balent, G. Goswami, V. Tomuta, D.T. Stein, Sensitive quantitative analysis of C-peptide in human plasma by 2-dimensional liquid chromatography-mass spectrometry isotope-dilution assay, Clin. Chem. 52 (2006) 872-879. https://doi.org/10.1373/clinchem.2005.063081

[34] E. Rogatsky, V. Tomuta, G. Cruikshank, L. Vele, H. Jayatillake, D. Stein, Direct sensitive quantitative LC/MS analysis of C-peptide from human urine by two dimensional reverse phase/reverse phase high-performance liquid chromatography, J. Sep. Sci. 29 (2006) 529-537. https://doi.org/10.1002/jssc.200500369

[35] A.V. Stoyanov, C. L. Rohlfing, S. Connolly, M. L. Roberts, C. L. Nauser, R.R. Little, Use of cation exchange chromatography for human C-peptide isotope dilution - Mass spectrometric assay, Journal of Chromatography. A 1218 (2011) 9244- 9249. https://doi.org/10.1016/j.chroma.2011.10.080

[36] X. Wang, L. Dai, Z. Q. Sun, P. Gao, Z. G. Ma, Separation and purification technology for total alkaloids from Uncariae Ramulus Cum Uncis with cation exchange resin, Chinese Traditional and Herbal Drugs. 42 (2011) 1973-1976.

[37] X. J. Peng, S. C. Li, Y. B. Li, H. Y. Ye, L. Yu, The Extraction, separation and identification of alkaloids in Leonurus heterophyllus, Research and Exploration in Laboratory. 33 (2014) 33-35.

[38] Q.Y. Fan, H. H. Zhang, W. H. Wang, Research on separation and purification of alkaloids in Cynoglossum amabile Stapf et Drumm with resin, Journal of Instrumental Analysis. 35 (2016) 1338-1342.

[39] J. J. Yan, Up-regulation on cytochromes P450 in Rat mediated by total alkaloid extract from Corydalis yanhusuo, BMC Complementary and Alternative Medicine. 14 (2014), Article No. 306. https://doi.org/10.1186/1472-6882-14-306

[40] Z. M. Sun, Z. G. Duan, L. Jiao, Z. F. Zhang, X. N. Li, Separation and Purification Process of Total Alkaloids from Sinisan, Chinese Journal of Experimental Traditional Medical Formulae. 17 (2011) 11-13.

[41] L. B. Guo, J. Z. Zhou, S. S. Zhu, Study on separation and purification of total alkaloids from Herba Ephedrae and Flos Daturae by macroporous adsorption resin, Food and Drug. 5 (2006) 47-49.

[42] M. Garcia-Vaquero, G. Rajauria, J.V. O'Doherty, T. Sweeney, Polysaccharides from macroalgae: recent advances, innovative technologies and challenges in extraction and purification, Food Res. Int. 99 (2017) 1011-1020. https://doi.org/10.1016/j.foodres.2016.11.016

[43] V. L. Campo, D. F. Kawano, D. B. d. Silva, I. Carvalho, Carrageenans: biological properties, chemical modifications and structural analysis - a review, Carbohydr. Polym. 77 (2009) 167-180. https://doi.org/10.1016/j.carbpol.2009.01.020

[44] Q. Cong, H. Chen, W. Liao, F. Xiao, P. Wang, Y. Qin, Q. Dong, K. Ding, Structural characterization and effect on anti-angiogenic activity of a fucoidan from Sargassum fusiforme, Carbohydr. Polym. 136 (2016) 899-907. https://doi.org/10.1016/j.carbpol.2015.09.087

[45] T. Imbs, S. Ermakova, O. Malyarenko, V. Isakov, T. Zvyagintseva, Structural elucidation of polysaccharide fractions from the brown alga Coccophora langsdorfii and in vitro investigation of their anticancer activity, Carbohydr. Polym. 135 (2016) 162-168. https://doi.org/10.1016/j.carbpol.2015.08.062

[46] A. Ramu Ganesan, M. Shanmugam, R. Bhat, Producing novel edible films from semi refined carrageenan (SRC) and ulvan polysaccharides for potential food applications, Int. J. Biol. Macromol. 112 (2018) 1164-1170. https://doi.org/10.1016/j.ijbiomac.2018.02.089

[47] M. Sterner, M. S. Ribeiro, F. Gröndahl, U. Edlund, Cyclic fractionation process for Saccharina latissima using aqueous chelator and ion exchange resin, J. Appl. Phycol. 29 (2017) 3175-3189. https://doi.org/10.1007/s10811-017-1176-5

Ion Exchange Resins: Biomedical and Environmental Applications   Materials Research Forum LLC
Materials Research Foundations 137 (2023) 93-119   https://doi.org/10.21741/9781644902219-6

Chapter 6

Ion Exchange Resins as Carriers for Sustained Drug Release

Bhavana Sampath Kumar, Junaiha Kapoor, Sandra Ravi M amd Dileep Francis*

Department of Life Sciences, Kristu Jayanti College, Autonomous, Bengaluru- 560077, Karnataka, India

* dileep@kristujayanti.com

Abstract

Ion exchange resins are water-insoluble cross-linked polymers conventionally used in chemical engineering to separate and purify substances from fluid mixtures and for the separation of gases. They are used in the pharmaceutical industry as a carrier for controlled drug delivery systems. Sustained drug release involves the release of the drug at a predetermined rate into the target system and helps maintain a constant drug concentration for a specified period. At present ion-exchange resins are widely studied and used as drug-delivery agents. Ion-exchange fibers are shought as an alternative due to their better efficiency in drug loading and delivery.

Keywords

Sustained Drug Delivery, Ion Exchange Resin, Resinate, Microencapsulation, Drug Release Kinetics

Contents

Ion Exchange Resins as Carriers for Sustained Drug Release 93

1.   Introduction 95

2.   Principles of sustained drug release 96

      2.1   Evolution of sustained drug delivery systems 97

      2.2.1 First-generation delivery systems 97

      2.2.2 Second-generation delivery systems 97

    2.2.3 Third/ Next generation delivery systems...98

3.   Types of sustained drug delivery systems 98

    3.1   Diffusion-controlled system 98

    3.1.1 Reservoir system..98

    3.1.2 Matrix system 98

    3.2   Osmotic system..99

    3.3   Floating system 99

    3.4   Bioadhesive system 99

    3.5   Liposome system 99

4.   IERs as drug delivery systems..100

    4.1   Chemistry of IERs 100

    4.2.   Complexation of IER and the drug.................................101

    4.2.1 Selection of the drug..102

    4.2.2 Purification of resins...102

    4.2.3 Drug loading 102

    4.2.3.1 Batch method 102

    4.2.3.2 Column method...102

    4.2.4 Factors affecting drug loading....................................103

    4.2.4.1 Particle size 103

    4.2.4.2 Porosity and swelling...104

    4.2.4.3 Available capacity...104

    4.2.4.4 Acid-base strength 104

    4.2.5 Evaluation of drug resinates 104

5.   Modified resinates...104

6.   Release kinetics of drugs complexed with IERs 106

7.   Efficiency of IERs as the delivery mechanism 107

    7.1   Oral drugs 107

    7.2   Nasal drugs 108

    7.3   Ophthalmic drugs..109

    7.4   Oro-dispersible films (ODF) 109

    7.5   Oral liquid suspensions..110

8.   Commercial IERs used in sustained drug delivery 110
        8.1   Dowex 50W 110
        8.2   Indion 244 110
        8.3   Amberlite IRP-69..111

9.   Future perspectives...111

References 112

1.   Introduction

Ion exchange resins (IERs) are small water-insoluble bead-like structures that are 1-2 mm in diameter. They are cross-linked polymers that can form complexes with chemicals, such as pharmaceutical drugs, due to the presence of ionizable functional groups [1]. Chemicals get trapped in the form of mobile ions on the resin and can be exchanged with the counterions present in the reaction medium. Based on the presence of positive or negative charges, they can be classified into anionic and cationic IERs, respectively [2]. IERs are widely used in wastewater treatment to remove contaminants like organic matter, heavy metals, dyes, and pharmaceutical waste [3]. They are also used in the pharmaceutical industry as a mechanism for taste masking; the taste of the pharmaceutical formulation plays an essential role in patient compliance [4]. Drugs that are bound to the IER are released only at a specific pH. Suppose the drug-resin complex is designed to exchange the drug with counterions in the gut pH. In that case, the drugs will not interact with the taste receptors in the oral cavity, and the patient will not experience the taste of the drug [5].

Another important pharmaceutical application of IER is in controlled drug delivery systems [6]. Before the use of IER, conventional methods such as dissolution and diffusion-controlled systems, osmotic systems, [7,8] floating systems, bioadhesive systems, and liposome systems were used for controlled drug delivery [9–11]. Owing to its advantages, the application of IER as a drug delivery system is now an active field of research. The drug, in its ionic form, is mixed with a suitable IER to form a complex known as the resinate. Drug-resin associations are mediated by ionic interactions between the immobilized functional group on the resin and the ionic charge on the drug. The resinate releases the drug as ions in exchange for the ionic species of the same charge, known as counter ions in the reaction medium. The rate of ion exchange depends upon the pH and counter ions present in the reaction medium. Since the gut pH and the counterions are known, IERs have become a logical choice for oral drug delivery systems. The pH dependence of the resinate in releasing drugs is utilized to ensure the sustained release of drugs in the gut [6].

2. Principles of sustained drug release

The therapeutically active component of a formulated drug is called the Active Pharmaceutical Ingredient (API). APIs are seldom administered directly into the target system due to a lack of control over the dosage and decreased stability of APIs when subjected to the ambient environment, incompatibility/ side effects at the administration site, and decreased bioavailability and unfavorable organoleptic properties [12]. In order to overcome these issues, APIs are formulated into dosage forms such as tablets, capsules, syrups, ointments, jellies, skin patches, and creams with the addition of a pharmaceutically inert excipient. Excipients give volume and shape to the drug, increase stability, modulate bioavailability, mask the unfavorable taste, and regulate release kinetics, amongst other advantages [13]. However, conventional drug formulations such as tablets and capsules do not regulate the release kinetics of the API. The total dosage is released rapidly into the system, resulting in poor absorption and bioavailability, premature metabolism and elimination, varying plasma concentrations, and the need for recurrent/higher dosage. Controlled drug delivery systems were developed in response to these challenges. Unlike conventional formulations, these systems follow zero-order kinetics and facilitate the release of a definite amount of API per unit of time [14].

Sustained drug delivery is a type of controlled drug delivery. However, it follows first-order kinetics because the proportion of drug released per unit of time is a function of drug concentration in the formulation. Nevertheless, drug release is controlled temporally. It involves the release of the drug at a predetermined rate into the system and maintaining a constant drug level for a specified period after a single dosage [15]. The development of administrative forms of drugs amenable to controlled or sustained delivery has become a keen interest in pharmaceutical research. They enable less frequent administration, improve therapeutic efficacy, reduce side effects, and facilitate patient convenience [16]. In the case of oral drugs that are immediately released into the gut, the concentration of the drug in the blood reaches its peak value immediately and then declines rapidly. This is not an optimal drug release strategy because the peak drug concentration is short-lived. The therapeutic effect wanes off soon after the release because of the rapid decline in concentration. A controlled drug delivery system ensures that the drug is released at a constant rate, which is important because the concentration of the drug is maintained in the blood with minimal fluctuations and within the therapeutically effective range for an extended period of time [17].

The release rates and the intensity and duration of drug activity can be controlled using physical and chemical systems. While physical systems utilize methods such as diffusion, erosion, and osmotic pressure, chemical systems utilize the formation of drug-polymer complex, drug-conjugate complex, prodrugs, or drug-resin complex. The drug release from the chemical systems is due to chemical reactions like hydrolysis, ion exchange, or enzymatic degradation [1].

2.1   Evolution of sustained drug delivery systems

The therapeutic activity of inhaled drugs used for respiratory diseases is limited because of their immediate clearance from the lungs. Hence, attempts were made to develop carriers for controlled drug delivery and retain the drug concentration in the lungs for a prolonged period by progressively releasing drugs at therapeutic levels. Substantial research has been done on the development of such carrier molecules, but not many drugs with this mechanism are available on the market [18].

Another class of therapeutics that benefits from sustained drug delivery is chemotherapy drugs. Conventional chemotherapy is often associated with severe side effects on non-target cells and tissues. An alternative would be to develop localized chemotherapy strategies wherein the drugs are targeted to the tumor site and released in a controlled manner. Such formulations may reduce the toxicity of conventional chemotherapy [19]. Shivakalyani et al., 2021, classified the controlled drug release systems developed over the years into three generations.

2.2.1   First-generation delivery systems

The first-generation controlled delivery strategies were developed from 1950 to 1980. These were designed mainly to enhance the efficiency of oral and transdermal drug formulations and dealt with small-molecule drugs. The four significant mechanisms adopted for drug release included dissolution, osmosis, diffusion, and ion exchange, and the former two were prevalent. A correlation between the in vitro drug release kinetics and in vivo pharmacokinetics for these formulations was established, and hence the clinical success rates were higher [20].

2.2.2   Second-generation delivery systems

These were developed between the 1990s to 2010. When compared to first-generation systems, clinical success rates of second-generation methods were lower. However, the delivery-associated problems addressed during this period were equally challenging. Controlled release of biopharmaceuticals such as peptides and proteins was attempted. Challenging administrative routes, such as the pulmonary route of delivery, were attempted due to a larger absorptive surface, thin membrane barriers, and robust blood supply. Biodegradable polymers, hydrogels, and liposomes were tried as delivery systems, especially for protein and peptide therapeutics. For instance, liposomal carriers were used for aerosol pulmonary delivery of insulin. However, these attempts were met with limited success because there was a lack of correlation between in vitro release kinetics and in vivo drug action. It was challenging to establish a predictive model for in vivo kinetics. Nanoparticle-mediated delivery of drugs was also attempted during this period [21].

2.2.3 Third/ Next generation delivery systems

These systems are envisaged as solutions to more challenging problems in drug -delivery. One ambitious goal is the self-regulation of the release of drugs in response to physiological requirements. One interesting example is a strategy to release insulin in response to blood glucose levels called Glucose Responsive Insulin Release (GRIR). Various approaches are attempted, including the incorporation of a subcutaneous insulin pump that can be modulated via a glucometer. Long-term delivery of biopharmaceuticals such as nucleic acids/ proteins/ peptides, delivery of poorly water-soluble drugs, and non-invasive administration of drugs are also priorities. Nanoparticles, innovative biopolymers, and hydrogels are the leading candidates for delivery agents [21,22].

3. Types of sustained drug delivery systems

3.1 Diffusion-controlled system

Diffusion controlled system involves two major strategies:

3.1.1 Reservoir system

These systems involve covering the drug with a water-insoluble polymer. Diffusion of the drug through the polymer is the rate-limiting step [23]. The polymer type and concentration determine the release kinetics of the drug. This system is not prevalent because sometimes clumping of the drug might occur due to non-uniform pore size in the polymer, which leads to potential toxicity. Moreover, it is challenging to deliver high molecular weight compounds [19].

3.1.2 Matrix system

Here, the delivery is mediated by a well-mixed composite of polymers with gelling properties. The composite enables the controlled release of the dissolved or dispersed drug in the system. The release rate depends on drug diffusion and not on dissolution. The system is cheap to formulate and is amenable to delivering high molecular weight compounds. However, the release rate can vary over time; it isn't easy to achieve zero-order kinetics with the system. Further, food ingredients and physiological factors may lead to dose dumping, and the GI transit time may affect release rates [7].

3.2   Osmotic system

This system involves using osmotic pumps in which the drug constitutes the inner core. A semipermeable membrane envelopes the core (osmogen). When the pump comes in contact with water or body fluids, liquid seeps into the core due to a change in osmotic pressure. The efflux leads to the expansion of core volume and pushes the drug solution through the delivery ports. The limiting factor of using the osmotic system as a sustained drug delivery mechanism is that a slight change in osmotic pressure or the rate of the solubility of the drug can affect the efficiency of the delivery system [8].

3.3   Floating system

A floating delivery system was formulated to prolong the residence of the drug in the gastric medium and improve bioavailability.   It was developed based on the principle of gas entrapment. The drug is coated with an effervescent agent (such as Sodium Carbonate) in the form of core pellets and subsequently encapsulated in a gas-entrapped polymeric coating. The system can stay afloat in the gut for around 24 hours and release the drug sustainably. However, the system requires a certain minimum level of gastric juices for the drug complexes to stay afloat [9,24].

3.4   Bioadhesive system

A prolonged residence time in the absorption site would help improve the drugs' pharmacokinetics and therapeutic efficacy. For instance, drugs absorbed through the gastrointestinal (GI) tract require increased residence time in the tract. Polymeric substances called bioadhesives are used as drug-delivery agents to achieve this goal. Bioadhesives attach to the surface of the mucosal membranes leading to sustained drug release and increasing the efficiency of the drug [17]. The major disadvantage of this mechanism is that the rate of influx of the drug is low due to which bioavailability of the drug reduces [10].

3.5   Liposome system

Liposomes are vesicles formed by the spontaneous aggregation of phospho- and sphingolipids. Inside cells, similar vesicles are used for the transport of biological macromolecules. Liposomes are widely used as drug delivery agents due to their biocompatibility and amphipathic properties. The phosphate groups make them partly polar, and the fatty acid chains are nonpolar. Liposomes are compatible with biological membranes and enable the delivery of drugs inside the cell. The kinetics and mode of delivery can be manipulated by modifying the composition of the liposomes. They may fuse with the cellular membrane or get endocytosed into the cells. They are used for carrying various therapeutics, such as small molecules, proteins, peptides, and antibodies. Liposomes have been used for sustained drug release at multiple sites, including the lungs and the eyes. However, conventionally formulated liposomes have a variety of disadvantages like low stability and shorter shelf life. Phospholipids are prone to hydrolysis and oxidation [11,25].

4.   IERs as drug delivery systems

IERs find applications as drug delivery agents for sustained drug release. Further, IERs enhance the drug's physical and chemical stability since the attached drug is more stable than the free drug. Weak ionic interactions mediate the association between the resin and the drug molecule. The drug molecule associated with the resin represents the fixed ion. The drug ions are released in exchange for a counter ion in the medium [26]. The affinity of the counterions and the drug ions towards the IER is competitive. At the delivery site, the resinate releases the drug by the ion exchange mechanism because of the presence of compatible counter ions [5]. Hence, the body serves as an exchanger for the drug ions. The ionic concentration of the biofluids in all individuals is almost similar; hence, the release rate of the drug is consistent in all individuals [1]. The naked IER without the drug molecule is eliminated from the delivery site [5].

4.1   Chemistry of IERs

IERs are made up of two principal components; a structural matrix and a functional component. The matrix is made up of a cross-linked polymer, such as polystyrene. A charged functional group is covalently attached to the matrix polymer. The functional component is the active ion species bound *via* ionic interactions to the functional group on the matrix. The matrix accounts for the physical properties of the IER and its interaction with biological substances. It also accounts for the drug-binding properties of IER [27].

The IER can be either a cation exchanger or an anion exchanger, depending on the charge of the resin. Cationic exchangers (IER^{-ve}) possess a negatively charged functional group that is covalently attached to the resin. Anionic exchangers have a positively charged functional group covalently attached to the resin [28].

The mechanism of drug exchange by cationic and anionic exchangers is given below:
Cationic exchanger: The fixed ion (Drug^{+ve}) bound to the resin will be positively charged. Positively charged counterions (A^{+ve}) in the medium will be exchanged with the fixed ion, as shown below:

$$IER^{-ve}\text{-}Drug^{+ve} + A^{+ve} \rightarrow Drug^{+ve} + IER^{-ve}\text{-}A^{+ve}$$

Anionic exchangers: They exchange the negatively charged fixed drug ions associated with them as shown below:

$$IER^{+ve}\text{- Drug}^{-ve} + B^{-ve} \rightarrow Drug^{-ve} + IER^{+ve}\text{-}B^{-ve}$$

Irrespective of the charge, the IER can further be classified as strong exchangers and weak exchangers based on the functional group attached. The strong IERs generally have sulphonic acid as functional groups, while the weak IERs contain carboxylic acid [26]. The mechanism of drug delivery by IERs in the case of oral drugs in the gastrointestinal (GI) tract is illustrated in **Figure 1.**

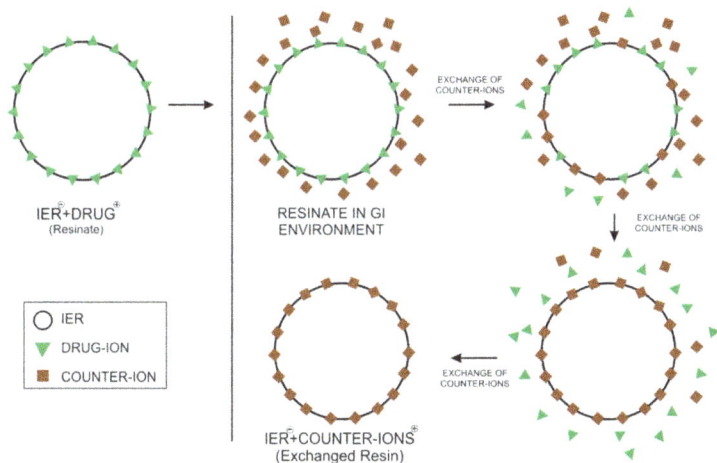

Figure 1. Mechanism of IER mediated drug delivery: A cationic resin, used to deliver a drug in the GI tract is shown.

4.2.   Complexation of IER and the drug

The proper complexation of the drug with the resinate is critical for effective drug delivery. The process involves various steps, and the success of the procedure depends on various parameters. The major steps in the formation of drug-resin complexes are depicted in **Figure 2.**

4.2.1 Selection of the drug

The drug should be charged and have either an acidic or basic functional group. The half-life of the API of the drug should be between 2-6 hours. The drug should be stable in the gastrointestinal fluid. Disintegration in the gut will lead to decreases therapeutic efficacy. The absorption zone of the drug in the gastrointestinal should be known. It should ideally be absorbed in all parts of the GI tract. The bioavailability of the drug is reduced if the absorption zone is limited to specific regions [1].

4.2.2 Purification of resins

Resins used for therapeutic delivery need to be purified before usage since they might contain impurities that might be toxic. Hence the first step in forming the resinate involves the purification of the resins. The cleansing of the resin can be done by two methods depending on the type of resin used. In the case of cationic resins, they are cycled repeatedly between sodium and hydrogen forms. In the case of anionic resins, they have to be cycled repeatedly between chloride and hydrogen forms. They are finally washed with water to remove any residual impurities and air-dried. For commercially used drugs, the size of the resin particles plays a significant role; hence the resins, after purification, are sieved to give a series of particle-size fragments [29].

4.2.3 Drug loading

Drug loading is usually performed using two methods:

4.2.3.1 Batch method

In the conventional batch method, the purified resin is dispersed in a known concentration of the drug solution and stirred till the drug-resin complex is formed, and an equilibrium is reached. Drug loading efficacy can be improved by using modified methods. In the double batch method, the supernatant formed after settling the drug-resin complex is decanted, and a fresh batch of the drug is added. The process is repeated once more in the triple batch method [1,30].

4.2.3.2 Column method

The resins are packed in a column through which a concentrated drug solution is passed until the eluent concentration is the same as the effluent concentration. A slurry of the resin is prepared in water and packed in a glass column. The column is stabilized by washing with water, and a known concentration of the drug is forced through at a specific flow rate. Eluent is collected at specific intervals, and the drug concentration is estimated [5,31].

After the formation of the resinate, it is filtered and washed. Depending upon the application, the resinate can either be used as a liquid suspension or as a solid dosage form. For liquid suspension applications, the filtrate is not dried. In the case of solid resinates, the filtrate is dried in a vacuum to form a powder that can be formulated into tablets [1].

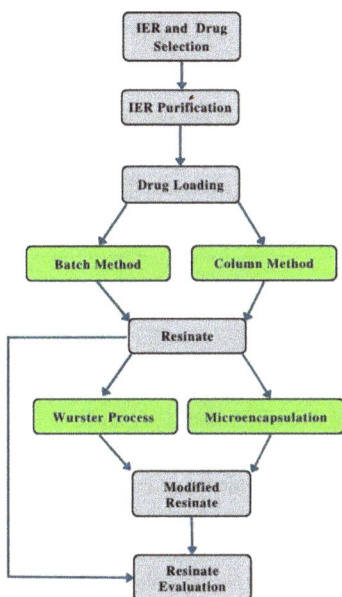

Figure 2. Steps in the production of drug-resin complexes.

4.2.4 Factors affecting drug loading

In both batch method and column methods, the time taken to reach equilibrium should be known for better drug loading. The factors that affect drug loading and release kinetics are as follows

4.2.4.1 Particle size

Particle size, the extent of particle cross-linking, and the nature of the resin will affect the time required to reach equilibrium [5]. Particle size is directly proportional to the time needed to attain equilibrium. With a smaller particle size, equilibrium is reached faster and hence using fine resin particles improves drug loading efficiency. Further, fine particles with lower surface area and smaller internal volume will reach equilibrium faster than coarse particles. Cross-linked IERs take more time to equilibrate with the drug than non-cross-linked resins [28].

4.2.4.2 Porosity and swelling

The number of cross-linking particles and the polymerization procedures affects the porosity of the resin. The porosity, in turn, affects the resin's swelling property, which affects the drug's release kinetics [30].

4.2.4.3 Available capacity

It is the quantitative measure of the ability of the resin to take up drug molecules as mobile ions, which are exchanged with the surrounding counter ions. The available capacity of the resinate depends on the functional group associated with the drug molecule [28].

4.2.4.4 Acid-base strength

The resinate's acid-base strength depends on the resins' ionic functional groups. Cation-exchange resins can have strong acid groups (sulphonic) or weak acid groups (carboxyl). The pK_a value of these resins ranges from 1 to 6. Anionic resins have a higher pKa value (9-13). The release of the drug in the gastric fluid is affected by the pK_a of the resin. Resins with strong acid groups were shown to release drugs more sustainably [30].

4.2.5 Evaluation of drug resinates

The amount of the drug in the resinate complex can be determined by the complete elution of the drug from the resin. The drugs are eluted from the resinate using a cationic exchanger such as hydrogen chloride (5-10% solution) or an anionic exchanger like sodium chloride (anionic exchanger), and the drug concentration in the eluent is determined. Complete elution becomes difficult in cases where the affinity between the drug and resin is strong; in such cases, an excess of competing counter ions is required for complete elution [1].

5. Modified resinates

Resinates in their pure form are convenient agents for sustained drug delivery systems. However, their drug release kinetics can be altered, and the release of the drugs can be delayed further by modifying IERs by applying an additional coating or extra layer. The coating serves as a diffusion barrier to the drug in the IER- drug complex and enhances the sustained release [32]. The mechanism of drug delivery using modified resonates is shown in **Figure 3**. A wide range of polymers can be used for the coating Ethylcellulose, for instance, is a commonly used cellulosic polymer marketed under the trade names such as

Aquacoat and Surerlease [33]. Eudragit is the market name for a polymethacrylate -based polymer widely used to coat IER- drug complexes [34]. Pharmaceutical waxes and carbohydrates are used as alternative coating materials. Due to its melting property, waxes can be maintained in the liquid state and coated over the resin by mixing them in a filtration tank. Once covered, the excess wax can be removed. The rate and strength of the coating depend on the polarity of the wax used. If they are highly polar, consisting of stearic acid and polyethylene glycol, they are bound tighter to the resinate than those with less polar groups like paraffin and stearyl alcohol [32].

Microencapsulation and Wurster Process (Fluidized Bed Process) are the two popular methods used for modifying resinates. Microencapsulation involves coating or depositing a membranous layer around the IER-drug complex with pharmaceutical–grade polymers to improve the drug release properties in the medium. In the reaction medium, the counter ions enter the membrane of the encapsulated resin and lead to the release of drug ions by diffusion through the membrane. The diffusion rate through the membranes depends on their composition and thickness. The rate of diffusion is inversely proportional to the thickness of the membrane [29].

Microencapsulation is preferred for preliminary studies or laboratory-scale production of drugs. Industry-scale manufacturing utilizes the Wurster Process or Fluidized Bed Process for coating. Wurster coaters characteristically contain a nozzle that sprays droplets of the coating material on the bottom of the fluidized bed. The particles flow cyclically through the fluidized bed, past the spray nozzle, and are deposited by the coating material. The coated materials move upwards to an expansion chamber with reduced air velocity and back to the coating chamber. The movement of the particles in the chamber ensures the evaporation of organic or aqueous solutions in which the coating material was dissolved, leaving only the coat material on the particle. Further, it prevents the clumping of particles [35].

Other major modified resinate systems involve:

Pennkinetic system: The IER-drug complexes can also be modified using the Pennkinetic system. Here, the resinates are first treated with polyethylene glycol, which improves the coating process and controls the rate of swelling of the resin. The pretreated resinates are subsequently coated with a water-insoluble polymer which regulates the diffusion pattern of the ions in and out of the system [5].

Gas-retentive system: The drugs which are primarily absorbed from the stomach should be stable for a prolonged period in the gastric environment. This can be achieved by using floating dosage forms [36]. A novel delivery system consisting of a bicarbonate resinate coated by a semipermeable membrane was used to improve the gastric residence time. The presence of chloride ions in the gastric tract, which is the counterion for carbonate, leads to the release of carbon dioxide, which is stored in the complex. This allows the resinate to float in the gastric environment and increases the gastric retention of the drug [37].

Sigmoidal release system: The release rate of the drug should be according to the therapeutic function and the nature of the active substance. In the sigmoidal release system, there is a predetermined lag phase during which the drug is released at a controlled rate, after which the drug is released from a multi-unit system at a rapid rate. Hence using this system, rhythmic and time-controlled release can be attained [38]. Usually, for sigmoidal release, a sugar bead is coated with an anionic exchange resin and covered by a layer consisting of a mixture of organic acids and drug ions. The organic acids are responsible for inducing an ionic environment that allows the drug's sigmoidal release [39].

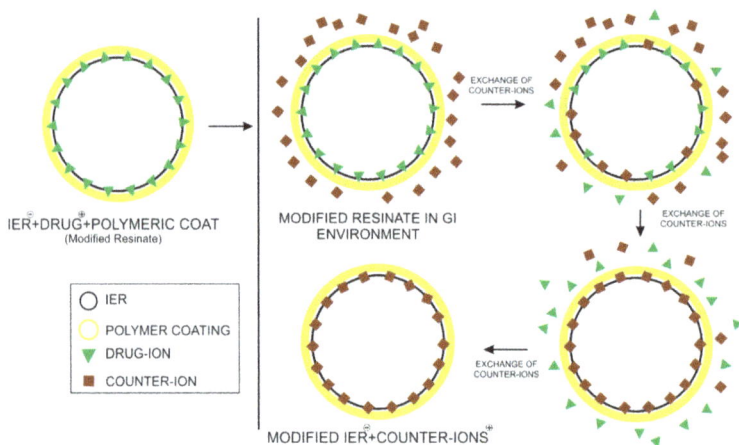

Figure 3. Mechanism of drug delivery by a modified resinate.

6.   Release kinetics of drugs complexed with IERs

The critical problems in drug delivery addressed by sustained drug-delivery systems are related to the drug release kinetics of conventional dosage forms. To be therapeutically effective in a system, the API should be available in circulation (plasma concentration of the drug) at an optimal level, between the minimum effective concentration (MEC) and toxic concentrations, for a specific window of time [40]. However, for most drugs,

Ion Exchange Resins: Biomedical and Environmental Applications   Materials Research Forum LLC
Materials Research Foundations 137 (2023) 93-119   https://doi.org/10.21741/9781644902219-6

especially those with a lower half-life, the availability of the drug in circulation at the optimal concentration is short-lived. Conventional dosage forms release the API in one shot and generally follow first-order kinetics. It results in an initial peaking of the drug plasma concentration followed by a rapid decline resulting in a lower residence time and reduced efficacy [41]. Multiple dosing, which can aggravate side effects and reduce patient compliance, helps to address the problem to some extent. Even then, the plasma levels of the drug fluctuate and will not remain constant. Sustained release systems such as IERs offer the best possible solution by maintaining the plasma concentration of the drug at a constant level for a prolonged period by following zero-order release kinetics. A fraction of the total API contained in the formulation is released initially to achieve the minimum effective concentration. Maintenance doses are released in a controlled manner periodically to compensate for drug elimination, thereby retaining the drug at the MEC for a prolonged period [42]. **Figure 4** illustrates the effect of various formulations on drug plasma concentrations as a function of time.

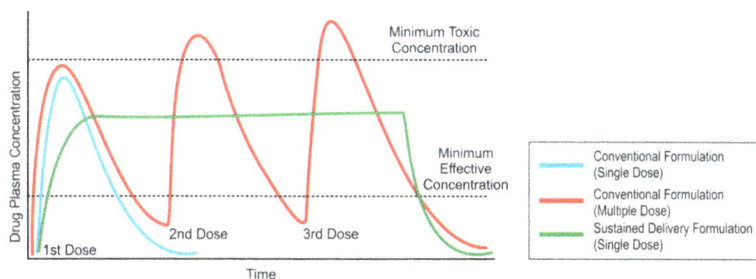

Figure 4. Changes in drug plasma concentrations against time for different drug formulations.

7.   Efficiency of IERs as the delivery mechanism

7.1   Oral drugs

Theophylline is a bronchodilator drug used to treat respiratory diseases such as chronic obstructive pulmonary disease and asthma [43]. Due to the various side effects caused by the drug, it is now used as an add-on therapy for respiratory diseases [44]. Theophylline formulated with an anionic IER and further modified by microencapsulation has shown better drug release properties [45].

Amoxicillin is used to treat peptic ulcers caused by *Helicobacter pylori*. Although the antibiotic is very effective against the bacterium under *in vitro conditions,* the effect is not reproduced *in vivo*, probably due to decreased gastric residence time and poor mucosal absorption [46]. Studies were conducted to target the drug to the gastric mucosa and predetermine its release rates. The drug was complexed with IER to form a resinate which was subsequently coated with a mucoadhesive layer. Evaluation of the gastric transit time in rats using fluorescence microscopy revealed that the coated drug had a higher residence time [47].

Dextromethorphan is an antitussive agent present in cold and cough medications [48]. A study demonstrated that the release rate of the drug could be controlled by increasing the size of the resin particle. The decrease in the release rate is due to a decrease in the diffusion coefficient and surface area. It was also found that the rate of release can further be reduced by coating the resin with an additional polymer layer [31].

Ranitidine hydrochloride is one of the most common oral drugs prescribed for heartburn [49]. Attempts were made to formulate this drug for sustained release to reduce dosage frequency and increase patient compliance. The drug was complexed with a strong cationic IER using a batch process. Formulation into the resinate resulted in higher bioavailability of the drug and resulted in sustained delivery [50].

7.2   Nasal drugs

A nasal drug formulation using IERs as a delivery system was developed for nicotine-replacement drugs. The resinate of nicotine and amberlite, in powder form, was shown to sustain the plasma nicotine levels at levels adequate for smoking cessation. The plasma profiles demonstrated that there was an initial surge in nicotine levels followed by a sustained elevated level. The concentration of nicotine during the sustained phase could be controlled by varying the ratio of free and bound nicotine [51].

In order to increase the efficiency of the intranasal immune response of the influenza hemagglutinin vaccine, which is a nanoparticle vaccine [52], it was administered along with sodium polystyrene sulfonate resins. It was observed that the resinate caused an increased level of mucosal IgA which led to a systemic increase in the hemagglutinin-inhibiting antibodies. It was also observed that the resinate could trigger a better immune response than the vaccine alone [53].

Nasal delivery of insulin was studied in rabbits. One study demonstrated that when human insulin complexed with an anionic resin (with optimum microparticle size) was administered through the nasal route. It led to an increase in insulin level after 15 mins and a decrease in glucose level after 45 mins showing higher bioavailability than insulin administered directly [54]. On the contrary, when the resinate was developed using a non-ionic resin or a cationic resin, similar results were not observed [55].

7.3 Ophthalmic drugs

Visual impairment is one of the primary reasons for the loss of productivity and diminishing quality of life for millions of people worldwide. Of the ophthalmic drugs available on the market, 70% are eye drops. However, it is observed that less than 5% of the drug is absorbed in the eye due to the existence of delivery barriers like blinking, nasolacrimal drainage, and the presence of cornea [56]. Hence, a mechanism that can increase the drug's prolonged residence and increase the drug's bioavailability becomes desirable. Formulations with materials like bioadhesive polymers, drug-loaded contact lenses, ointments, IERs, and preformed gels were studied for the process. It was demonstrated that IERs are the best candidate material for prolonging the residence of the drug in the eye [57].

Ciprofloxacin hydrochloride is another drug prescribed for the treatment of bacterial conjunctivitis [58]. The conventional drug required multiple doses per day; hence attempts were made to formulate the drug into forms that would reduce the dosage to once a day for better patient compliance. The drugs were complexed with IER and coated with polymers. The resinate formed was stable and resulted in sustained *in vitro* drug release for about 24 hours [59].

7.4 Oro-dispersible films (ODF)

Betahistine hydrochloride is an oral drug used to treat vertigo and balance disorders, available in the form of tablets. Since the incidence of these diseases is unpredictable and could be severe, immediate drug administration may be required. However, the availability of water to ingest the drug is not always guaranteed. Further, drinking water can increase the misery of the patients [60]. Hence the drug is taken in the form of ODFs, which are drugs embedded within a thin film. The film is made of substances that are readily dissolved in the oral cavity. The drug is dissolved in saliva and eventually gets absorbed in the GI tract. Hence the drugs with ODF can be administered without water. However, betahistine hydrochloride is a fast-disintegrating drug and requires frequent administration. Mixing the API with IERs and formulating it into ODFs resulted in a sustained release compared to the ODFs with the naked API. Another advantage of complexing the drug with IER is that it masks the taste and increases patient compliance [61].

7.5   Oral liquid suspensions

Clonidine hydrochloride (CH) is a drug used to treat hypertension. The dosage forms of CH available in the market include quick-release tablets and patches. These formulations are not ideal for elderly patients due to a lack of uniformity in dose distribution, the absence of a flexible method of administration, and the need for frequent administration owing to the sudden decline of the drug concentration in the blood [62]. Hence attempts have been made to develop alternative dosage forms like liquid suspensions and sustained-release forms. Many studies have demonstrated success with IERs as a delivery mechanism for sustained drug release [63].

Allergic rhinitis is a chronic inflammatory disorder seen in childhood and adolescence. The most common symptoms of this disease are sneezing nasal congestion, and rhinorrhea [64]. The most common treatment is the administration of antihistamines as an oral drug. Carbinoxamine maleate (CAM) is an antihistamine widely prescribed for a variety of allergic reactions. Though CAM in its oral dosage form has been in clinical use for many years, it is not devoid of shortcomings. It has a very short half-life necessitating frequent dose administration (3 - 4 times a day). It can also cause gastrointestinal irritation and have a bitter taste [65]. Generally, children are affected by allergic rhinitis; hence the drug is preferred in the form of an oral liquid suspension for better compliance. Suspensions of microencapsulated CAM complexed with IERs were demonstrated to be stable and achieved sustained release *in vivo*. Further, the dosage form is suitable for administration in pediatric patients and effectively masks the bitter taste [66].

8.   Commercial IERs used in sustained drug delivery

8.1   Dowex 50W

Dowex 50W is a strong cationic synthetic IER widely used for softening water and in sustained drug delivery. The matrix of the resin is made up of styrene-divinylbenzene and is functionalized by Sulfonic acid. Diltiazem, a calcium channel blocker used in the treatment of hypertension and chest ache, has a short half-life and hence requires frequent dosage. Diltiazem complexed with Dowex 50W resulted in the sustained release of the drug [67].

8.2   Indion 244

Indion 244 is a strong synthetic cationic exchange resin used widely in pharmaceutical formulations. The sustained release of chlorpheniramine maleate, an antihistamine, was achieved by complexing it with Indion 244 [68]. Metformin HCL, an antidiabetic medication, was complexed with Indion 244 to achieve sustained drug release under in vitro conditions [69].

8.3   Amberlite IRP-69

Amberlite IRP-69 is a synthetic cationic resin widely used for taste masking and sustained delivery of drugs. Famotidine, a histamine H_2 receptor antagonist, is used for the treatment of heartburn. The efficiency of the drug is hampered by its bitter taste and short half-life. The drug was associated with Amberlite IRP-69 to form a resinate. The formulation increased the bioavailability of the drug and successfully masked its taste [70].

9.   Future perspectives

A large number of studies have been conducted to develop a delivery mechanism suitable for sustained drug delivery over the past two decades. However, current approaches are not fool proof, and there is enormous scope for improvement and innovations. Sustained drug delivery is proven better than conventional drug delivery as it is patient-friendly. Still, the present technologies used for sustained drug delivery have their own demerits. The prolonged residence time in the system achieved through sustained delivery may lead to increased side effects. Further, If a drug is proven toxic after administration, removal of the drug from the system is rendered difficult [71]. The drug release kinetics observed for the drugs complexed with IERs seldom show a correlation under *in vitro and in vivo* conditions

Although IERs are widely used as a drug delivery mechanism, attempts are being made to replace IERs with ion-exchange fibers. Ion exchange fibers are similar to IERs except that IERs are cross-linked while ion exchange fibers are not cross-linked [72]. Various studies have demonstrated that ion-exchange fibers possess better drug-loading ability, have better control of ion-exchange mechanisms, and are helpful in the better incorporation of drug-sized compounds. It has also been observed that ion-exchange fibers have better chemical, mechanical and thermal stability than IERs [73].

In general, transdermal iontophoretic drug delivery (delivery of a drug across the skin by applying a small electric potential) has very low efficiency. It is because the electric current contributed by the drug ions is very low. In order to enhance the contribution of the drug to the total electric current, the drug Diclofenac sodium was formulated with various materials such as IERs, ion exchange membranes, and ion exchange fibers, and the rate of transdermal drug delivery across rat skin was determined. It was demonstrated that among the different conjugates, ion exchange fibers contributed to the highest increase in the rate of transdermal delivery. Hence ion-exchange fibers are now being studied as a delivery mechanism for transdermal drugs [74].

Exchange fibers have shown promise as a sustained delivery agent in many oral drug suspensions. The use of Propranolol hydrochloride (PPN), a drug usually used to treat hypertension and cardiovascular disorders, is hampered by its short half-life and bitter taste. The drug was formulated with ion-exchange fibers and coated with a polymer to mask the taste and prolong the release. Fiber-drug complex and coated fiber-drug complex were stable and resulted in sustained release of the drug. Further, there was a correlation between in vitro dissolution and in vivo absorption [75]. Hence, future formulations for the sustained delivery of drugs may rely more on fibers rather than resins.

References

[1] X. Guo, R.K. Chang, M.A. Hussain, Ion-exchange resins as drug delivery carriers, J. Pharm. Sci. 98 (2009) 3886-3902. https://doi.org/10.1002/jps.21706

[2] Z. Liu, E.C. Mills, M. Mohseni, B. Barbeau, P.R. Bérubé, Biological ion exchange as an alternative to biological activated carbon for natural organic matter removal: Impact of temperature and empty bed contact time (EBCT), Chemosphere. 288 (2022) 132466. https://doi.org/10.1016/j.chemosphere.2021.132466

[3] Z. Liu, M. Haddad, S. Sauvé, B. Barbeau, Alleviating the burden of ion exchange brine in water treatment: From operational strategies to brine management, Water Res. 205 (2021) 117728. https://doi.org/10.1016/j.watres.2021.117728

[4] M.T.R. Chikukwa, M. Wesoly, A.B. Korzeniowska, P. Ciosek-Skibinska, R.B. Walker, S.M.M. Khamanga, Assessment of taste masking of captopril by ion-exchange resins using electronic gustatory system, Pharm. Dev. Technol. 25 (2020) 281-289. https://doi.org/10.1080/10837450.2019.1687520

[5] V. Anand, R. Kandarapu, S. Garg, Ion-exchange resins: Carrying drug delivery forward, Drug Discov. Today. 6 (2001) 905-914. https://doi.org/10.1016/S1359-6446(01)01922-5

[6] HJ Seong, NH. Berhane, K. Haghighi, K. Park, Drug release properties of polymer coated ion-exchange resin complexes: Experimental and theoretical evaluation, J. Pharm. Sci. 96 (2007) 618-632. https://doi.org/10.1002/jps.20677

[7] J. Siepmann, R.A. Siegel, F. Siepmann, Diffusion Controlled Drug Delivery Systems, in: J. Siepmann, R. Siegel, M. Rathbone, (Eds.) Fundamentals and Applications of Controlled Release Drug Delivery. Advances in Delivery Science and Technology, Springer, Boston, MA., 2012, pp. 388-398. https://doi.org/10.1007/978-1-4614-0881-9

Materials Research Foundations 137 (2023) 93-119   https://doi.org/10.21741/9781644902219-6

[8] CN Patra, S. Swain, J. Sruti, AP Patro, K.C. Panigrahi, S. Beg, M.E.B. Rao, Osmotic drug delivery systems: basics and design approaches, Recent Pat. Drug Deliv. Formul. 7 (2013) 150-161. https://doi.org/10.2174/1872211311307020007

[9] B. Kumari, A. Khansili, P. Phougat, M. Kumar, Comprehensive review of the role of acrylic acid derivative polymers in floating drug delivery systems, Polim. Med. 49 (2019) 71-79. https://doi.org/10.17219/pim/122016

[10] K. Kumar, N. Dhawan, H. Sharma, S. Vaidya, B. Vaidya, Bioadhesive polymers: a novel tool for drug delivery, Artif. Cells, Nanomedicine, Biotechnol. 42 (2014) 274-283. https://doi.org/10.3109/21691401.2013.815194

[11] Z.M. Hu, S.Y. Liu, H.Y. Yang, C. Huang, Research progress of liposome drug delivery system in stomatology, Chin. J. Stomatol. 56 (2021) 294-300.

[12] S. Adepu, S. Ramakrishna, Controlled Drug Delivery Systems: Current Status and Future Directions, Molecules. 26 (19) (2021) 5905. https://doi.org/10.3390/molecules26195905

[13] V. Kumar, V. Bansal, A. Madhavan, M. Kumar, R. Sindhu, M. K. Awasthi, P. Binod, S. Saran, Active pharmaceutical ingredient (API) chemicals: a critical review of current biotechnological approaches, Bioengineered. 13(2)(2022) 4309-4327. https://doi.org/10.1080/21655979.2022.2031412

[14] Y. Zhang, H.F. Chan, K.W. Leong, Advanced materials and processing for drug delivery: The past and the future, Adv. Drug Deliv. Rev. 65 (2013) 104-120. https://doi.org/10.1016/j.addr.2012.10.003

[15] K.R. Reddy, S. Mutalik, S. Reddy, Once-daily sustained-release matrix tablets of nicorandil: Formulation and in vitro evaluation, AAPS PharmSciTech. 4 (2003) 480-488. https://doi.org/10.1208/pt040461

[16] B. Sun, M. Zhang, J. Shen, Z. He, P. Fatehi, Y. Ni, Applications of cellulose-based materials in sustained drug delivery systems, Curr. Med. Chem. 26 (2018) 2485-2501. https://doi.org/10.2174/0929867324666170705143308

[17] H. Kojima, K. Yoshihara, T. Sawada, H. Kondo, K. Sako, Extended release of a large amount of highly water-soluble diltiazem hydrochloride by utilizing counter polymer in polyethylene oxides (PEO)/polyethylene glycol (PEG) matrix tablets, Eur. J. Pharm. Biopharm. 70 (2008) 556-562. https://doi.org/10.1016/j.ejpb.2008.05.032

[18] C. Loira-Pastoriza, J. Todoroff, R. Vanbever, Delivery strategies for sustained drug release in the lungs, Adv. Drug Deliv. Rev. 75 (2014) 81-91. https://doi.org/10.1016/j.addr.2014.05.017

[19] M. Norouzi, B. Nazari, D.W. Miller, Injectable hydrogel-based drug delivery systems for local cancer therapy, Drug Discov. Today. 21 (2016) 1835-1849. https://doi.org/10.1016/j.drudis.2016.07.006

[20] Y.H. Yun, B.K. Lee, K. Park, Controlled Drug Delivery: Historical perspective for the next generation, J. Control. Release. 219 (2015) 2-7. https://doi.org/10.1016/j.jconrel.2015.10.005

[21] S. Adepu, S. Ramakrishna, Controlled Drug Delivery Systems: Current Status and Future Directions, Molecules. 26 (2021) 5905. https://doi.org/10.3390/molecules26195905

[22] K. Park, The Controlled Drug Delivery Systems: Past Forward and Future Back, J. Control. Release. 190 (2014) 3-8. https://doi.org/10.1016/j.jconrel.2014.03.054

[23] R.S. Langer, N.A. Peppas, Present and future applications of biomaterials in controlled drug delivery systems, Biomaterials. 2 (1981) 201-214. https://doi.org/10.1016/0142-9612(81)90059-4

[24] S. Sungthongjeen, O. Paeratakul, S. Limmatvapirat, S. Puttipipatkhachorn, Preparation and in vitro evaluation of a multiple-unit floating drug delivery system based on gas formation technique, Int. J. Pharm. 324 (2006) 136-143. https://doi.org/10.1016/j.ijpharm.2006.06.002

[25] D.K. Karumanchi, Y. Skrypai, A. Thomas, E.R. Gaillard, Rational design of liposomes for sustained release drug delivery of bevacizumab to treat ocular angiogenesis, J. Drug Deliv. Sci. Technol. 47 (2018) 275-282. https://doi.org/10.1016/j.jddst.2018.07.003

[26] D. Mastropietro, K. Park, H. Omidian, Polymers in Oral Drug Delivery, in P. Ducheyne (Ed.), Comprehensive Biomaterials II, Elsevier, Oxford, (2017) 430-444. https://doi.org/10.1016/B978-0-12-803581-8.09291-2

[27] J. Brady, T. Dürig, P. I. Lee, J,-X, Li, Polymer Properties and Characterization, in Y. Qiu, Y. Chen, G. G. Z. Zhang, L. Yu, R. V. Mantri (Eds.), Developing Solid Oral Dosage Forms (Second Edition), Academic Press, Boston, (2017) 181-223. https://doi.org/10.1016/B978-0-12-802447-8.00007-8

[28] S. D. Alexandratos. Ion-Exchange Resins: A Retrospective from Industrial and Engineering Chemistry Research, Ind. Eng. Chem. Res. 48 (2009) 388-398. https://doi.org/10.1021/ie801242v

[29] D. Torres, B. Seijo, G. García-Encina, M.J. Alonso, J.L. Vila-Jato, Microencapsulation of ion-exchange resins by interfacial nylon polymerization, Int. J. Pharm. 59 (1990) 9-17. https://doi.org/10.1016/0378-5173(90)90059-D

[30] A.B. Jumde, M.J. Umekar, N.R. Kotagale, Complexation using direct current: novel batch method for drug-resinate preparation, Drug Dev. Ind. Pharm. 39 (2013) 978-984. https://doi.org/10.3109/03639045.2012.692375

[31] S.H. Jeong, K. Park, Drug loading and release properties of ion-exchange resin complexes as a drug delivery matrix, Int. J. Pharm. 361 (2008) 26-32. https://doi.org/10.1016/j.ijpharm.2008.05.006

[32] S.H. Jeong, K. Park, Simple preparation of coated resin complexes and their incorporation into fast-disintegrating tablets, Arch. Pharm. Res. 33 (2010) 115-123. https://doi.org/10.1007/s12272-010-2233-7

[33] M. Gay Moldenhauer, J. Graham Nairn, Formulation parameters affecting the preparation and properties of microencapsulated ion-exchange resins containing theophylline, J. Pharm. Sci. 79 (1990) 659-666. https://doi.org/10.1002/jps.2600790802

[34] M.C. Adeyeye, E. Mwangi, S. Katpally, K. Fujioka, H. Ichikawa, Y. Fukumori, Suspensions of prolonged-release diclofenac-Eudragit® and ion-exchange resin microcapsules: II. Improved dissolution stability, J. Microencapsul. 22 (2005) 353-362. https://doi.org/10.1080/02652040500100865

[35] V. Mohylyuk, K. Patel, N. Scott, C. Richardson, D. Murnane, F. Liu, Wurster fluidized bed coating of microparticles: towards scalable production of oral sustained-release liquid medicines for patients with swallowing difficulties, AAPS PharmSciTech. 21 (2019). https://doi.org/10.1208/s12249-019-1534-5

[36] F. Atyabi, H.L. Sharma, H.A.H. Mohammad, J.T. Fell, Controlled drug release from coated floating ion exchange resin beads, J. Control. Release. 42 (1996) 25-28. https://doi.org/10.1016/0168-3659(96)01343-0

[37] R.B. Umamaheshwari, S. Jain, N.K. Jain, A new approach in gastroretentive drug delivery system using cholestyramine, Drug Deliv. 10 (2003) 151-160. https://doi.org/10.1080/713840399

[38] S. Narisawa, M. Nagata, Y. Hirakawa, M. Kobayashi, H. Yoshino, An organic acid-induced sigmoidal release system for oral controlled-release preparations. 2. Permeability enhancement of Eudragit RS coating led by the physicochemical

interactions with organic acid, J. Pharm. Sci. 85 (1996) 184-188. https://doi.org/10.1021/js950180o

[39] S. Narisawa, M. Nagata, C. Danyoshi, H. Yoshino, K. Murata, Y. Hirakawa, K. Noda, An organic acid-induced sigmoidal release system for oral controlled-release preparations, Pharm. Res. 11 (1994) 111-116. https://doi.org/10.1023/A:1018910114436

[40] J. Tamargo, J.Y. Le Heuzey, P. Mabo, Narrow therapeutic index drugs: a clinical pharmacological consideration to flecainide, Eur. J. Clin. Pharmacol. 71 (2015) 549-567. https://doi.org/10.1007/s00228-015-1832-0

[41] G. Yadav, M. Bansal, N. Thakur, Sargam, P. Khare, Multilayer tablets and their drug release kinetic models for oral controlled drug delivery systems, Middle East J. Sci. Res. 16 (2013) 782-795.

[42] G. Yadav, M. Bansal, N. Thakur, P. Khare, Multilayer Tablets and Their Drug Release Kinetic Models for Oral Controlled Drug Delivery System, Middle-East J. Sci. Res. 16 (2013) 782-795.

[43] R. Wettengel, Theophylline--past present and future, Arzneimittelforschung. 48 (1998) 535-539.

[44] R.I. Ogilvie, Monitoring plasma theophylline concentrations, Ther. Drug Monit. 2 (1980) 111-117. https://doi.org/10.1097/00007691-198004000-00001

[45] S. Salatin, J. Barar, M. Barzegar-Jalali, K. Adibkia, M. Alami-Milani, M. Jelvehgari, Formulation and evaluation of Eudragit RL-100 nanoparticles loaded in-situ forming gel for intranasal delivery of Rivastigmine, Adv. Pharm. Bull. 10 (2020) 20. https://doi.org/10.15171/apb.2020.003

[46] R.N. Brogden, R.C. Heel, T.M. Speight, G.S. Avery, Amoxicillin injectable: a review of its antibacterial spectrum, pharmacokinetics and therapeutic use, Drugs. 18 (1979) 169-184. https://doi.org/10.2165/00003495-197918030-00001

[47] Cuna, M.J. Alonso, D. Torres, Preparation and in vivo evaluation of mucoadhesive microparticles containing amoxycillin-resin complexes for drug delivery to the gastric mucosa, Eur. J. Pharm. Biopharm. 51 (2001) 199-205. https://doi.org/10.1016/S0939-6411(01)00124-2

[48] A.R. Silva, R.J. Dinis-Oliveira, Pharmacokinetics and pharmacodynamics of dextromethorphan: clinical and forensic aspects, Drug Metab. Rev. 52 (2020) 258-282. https://doi.org/10.1080/03602532.2020.1758712

[49] T.S. Gaginella, J.H. Bauman, Ranitidine hydrochloride, Drug Intell. Clin. Pharm. 17 (1983) 873-885. https://doi.org/10.1177/106002808301701201

[50] S. Khan, A. Guha, P. Yeole, P. Katariya, Strong cation exchange resin for improving physicochemical properties and sustaining release of ranitidine hydrochloride, Indian J. Pharm. Sci. 69 (2007) 626. https://doi.org/10.4103/0250-474X.38466

[51] Y.H. Cheng, P. Watts, M. Hinchcliffe, R. Hotchkiss, R. Nankervis, N.F. Faraj, A. Smith, S.S. Davis, L. Illum, Development of a novel nasal nicotine formulation comprising an optimal pulsatile and sustained plasma nicotine profile for smoking cessation, J. Control. Release. 79 (2002) 243-254. https://doi.org/10.1016/S0168-3659(01)00553-3

[52] H. Jeong, C.S. Lee, J. Lee, J. Lee, H.S. Hwang, M. Lee, K. Na, Hemagglutinin nanoparticulate vaccine with controlled photochemical immunomodulation for pathogenic influenza-specific immunity, Adv. Sci. Weinheim, Baden-Wurttemberg, Ger. 8 (2021). https://doi.org/10.1002/advs.202100118

[53] M. Higaki, T. Takase, R. Igarashi, Y. Suzuki, C. Aizawa, Y. Mizushima, Enhancement of immune response to intranasal influenza HA vaccine by microparticle resin, Vaccine. 16 (1998) 741-745. https://doi.org/10.1016/S0264-410X(97)00248-X

[54] M. Takenaga, Y. Serizawa, Y. Azechi, A. Ochiai, Y. Kosaka, R. Igarashi, Y. Mizushima, Microparticle resins as a potential nasal drug delivery system for insulin, J. Control. Release. 52 (1998) 81-87. https://doi.org/10.1016/S0168-3659(97)00193-4

[55] J. Wang, Y. Tabata, K. Morimoto, Aminated gelatin microspheres as a nasal delivery system for peptide drugs: evaluation of in vitro release and in vivo insulin absorption in rats, J. Control. Release. 113 (2006) 31-37. https://doi.org/10.1016/j.jconrel.2006.03.011

[56] V. Gote, S. Sikder, J. Sicotte, D. Pal, Ocular Drug Delivery: Present Innovations and Future Challenges, J. Pharmacol. Exp. Ther. 370 (2019) 602-624. https://doi.org/10.1124/jpet.119.256933

[57] Y. Wei, C. Li, Q. Zhu, X. Zhang, J. Guan, S. Mao, Comparison of thermosensitive in situ gels and drug-resin complex for ocular drug delivery: In vitro drug release and in vivo tissue distribution, Int. J. Pharm. 578 (2020) 119184. https://doi.org/10.1016/j.ijpharm.2020.119184

[58] D.K. Terp, M.J. Rybak, Ciprofloxacin, Drug Intell. Clin. Pharm. 21 (1987) 568-574. https://doi.org/10.1177/1060028087021007-801

[59] S.P. Jain, S.P. Shah, N.S. Rajadhyaksha, PSPS Pirthi, P.D. Amin, In situ ophthalmic gel of ciprofloxacin hydrochloride for once a day sustained delivery, Drug Dev. Ind. Pharm. 34 (2008) 445-452. https://doi.org/10.1080/03639040701831710

[60] R. Ramos Alcocer, J.G. Ledezma Rodríguez, A. Navas Romero, J.L. Cardenas Nuñez, V. Rodríguez Montoya, J. Deschamps, J.A. Liviac Ticse, Use of betahistine in the treatment of peripheral vertigo, Acta Otolaryngol. 135 (2015) 1205-1211. https://doi.org/10.3109/00016489.2015.1072873

[61] R. Shang, C. Liu, P. Quan, H. Zhao, L. Fang, Effect of drug-ion exchange resin complex in betahistine hydrochloride orodispersible film on sustained release, taste masking and hygroscopicity reduction, Int. J. Pharm. 545 (2018) 163-169. https://doi.org/10.1016/j.ijpharm.2018.05.004

[62] M.C. Houston, Clonidine hydrochloride, South. Med. J. 75 (1982) 713-721. https://doi.org/10.1097/00007611-198206000-00022

[63] H. Liu, X. Xie, C. Chen, C.K. Firempong, Y. Feng, L. Zhao, X. Yin, Preparation and in vitro/in vivo evaluation of a clonidine hydrochloride drug-resin suspension as a sustained-release formulation, Drug Dev. Ind. Pharm. 47 (2021) 394-402. https://doi.org/10.1080/03639045.2021.1890110

[64] C.F. Schuler IV, J.M. Montejo, Allergic rhinitis in children and adolescents, Pediatr. Clin. North Am. 66 (2019) 981-993. https://doi.org/10.1016/j.pcl.2019.06.004

[65] Y. Liu, P. Li, R. Qian, T. Sun, F. Fang, Z. Wang, X. Ke, B. Xu, A novel and discriminative method of in vitro disintegration time for preparation and optimization of taste-masked orally disintegrating tablets of carbinoxamine maleate, Drug Dev. Ind. Pharm. 44 (2018) 1317-1327. https://doi.org/10.1080/03639045.2018.1449854

[66] Y. Deng, T. Wang, J. Li, W. Sun, H. He, J. Gou, Y. Wang, T. Yin, Y. Zhang, X. Tang, Studies on the in vitro ion exchange kinetics and thermodynamics and in vivo pharmacokinetics of the carbinoxamine-resin complex, Int. J. Pharm. 588 (2020). https://doi.org/10.1016/j.ijpharm.2020.119779

[67] V.B. Junyaprasert, G. Manwiwattanakul, Release profile comparison and stability of diltiazem-resin microcapsules in sustained release suspensions, Int. J. Pharm. 352 (2008) 81-91. https://doi.org/10.1016/j.ijpharm.2007.10.018

[68] A. Kadam, D. Sakarkar, P. Kawtikwar, Development and Evaluation of Oral Controlled release chlorpheniramine-ion exchange resinate suspension, Indian J. Pharm. Sci. 70 (2008) 531. https://doi.org/10.4103/0250-474X.44613

[69] P.K. Bhoyar, D.M. Biyani, Formulation and in vitro evaluation of sustained release dosage form with taste masking of metformin hydrochloride, Indian J. Pharm. Sci. 72 (2010) 184. https://doi.org/10.4103/0250-474X.65031

[70] RM okhta. Aman, M.M. ohame. Meshali, G.M. ahmou. Abdelghani, Ion-exchange complex of famotidine: sustained release and taste masking approach of stable liquid dosage form, Drug Discov. Ther. 8 (2014) 268-275. https://doi.org/10.5582/ddt.2014.01043

[71] A. Minocha, D.A. Spyker, Acute overdose with sustained release drug formulations. Perspectives in treatment, Med. Toxicol. 1 (1986) 300-307. https://doi.org/10.1007/BF03259845

[72] J. Yuan, Y. Gao, X. Wang, H. Liu, X. Che, L. Xu, Y. Yang, Q. Wang, Y. Wang, S. Li, The load and release characteristics on a strong cationic ion-exchange fiber: Kinetics, thermodynamics, and influences, Drug Des. Devel. Ther. 8 (2014) 945-955. https://doi.org/10.2147/DDDT.S64604

[73] M. Vuorio, J.A. Manzanares, L. Murtomäki, J. Hirvonen, T. Kankkunen, K. Kontturi, Ion-exchange fibers and drugs: A transient study, J. Control. Release. 91 (2003) 439-448. https://doi.org/10.1016/S0168-3659(03)00270-0

[74] C. Xin, W. Li-Hong, Y. Yue, G. Ya-Nan, W. Qi-Fang, Y. Yang, L. San-Ming, A novel method to enhance the efficiency of drug transdermal iontophoresis delivery by using complexes of drug and ion-exchange fibers, Int. J. Pharm. 428 (2012) 68-75. https://doi.org/10.1016/j.ijpharm.2012.02.039

[75] J. Yuan, T. Liu, H. Li, T. Shi, J. Xu, H. Liu, Z. Wang, Q. Wang, L. Xu, Y. Wang, S. Li, Oral sustained-release suspension based on a novel taste-masked and mucoadhesive carrier-ion-exchange fiber, Int. J. Pharm. 472 (2014) 74-81. https://doi.org/10.1016/j.ijpharm.2014.05.048

Ion Exchange Resins: Biomedical and Environmental Applications   Materials Research Forum LLC
Materials Research Foundations 137 (2023) 120-141   https://doi.org/10.21741/9781644902219-7

Chapter 7

Ion Exchange Resins for Clinical Applications

Muhammad Hassan Sarfraz[1], Mohsin Khurshid[1], Bilal Aslam[1], Muhammad Asif Zahoor[1], Sumreen Hayat[1], Muhammad Saqalein[1], Muhammad Farrukh Sarfraz[2], Saima Muzammil[1]*

[1]Departmant of Microbiology, Government College University Faisalabad, Allama Iqbal road, Faisalabad, Pakistan

[2]Department of Physics, COMSATS University Islamabad, Lahore Campus, Pakistan

* saimamuzammil83@gmail.com

Abstract

Ion exchange resins are the cross-linked polymers that are insoluble in water and consist of charged ionic groups, for exchanging counter-ions, at repeating positions along the resin. IERs are majorly classified into two types i.e., cationic exchange resins and anionic exchange resins, depending on the type of ion exchanged. Over the last few years, ion exchange resins have gained significant attention owing to their various benefits in drug formulations improving the taste, stability, anti-deliquescence, and dissolution of the drug. They also play a role in targeted drug delivery via employing various delivery technologies such as sigmoidal release system, gastric retentive system, hollow fiber system and microencapsulated resonates. Furthermore, they are also used for therapeutic purposes like treating different ailments and abnormalities such as treating high cholesterol, edema, pruritus, cardiac edema, uremia etc. The chapter presents an overview of the role of ion exchange resins in various clinical implications.

Keywords

Cationic Resins, Anionic Resins, Drug Release, Targeted Drug Delivery, Therapeutics

Contents

Ion Exchange Resins for Clinical Applications..120

1.   Introduction..122

2.   Application of resins in formulation-related issues............................123

2.1   Taste development 123

2.2   Aiding in dissolution 124

2.3   Role as disintegrating agents 125

2.4   Drug stabilization 125

2.5   Water purification for the production of pharmaceuticals 126

2.6   Anti-deliquescence 126

3.   **Applications in drug release systems** **126**

3.1   Simple resinates 128

3.2   Microencapsulated resinates 128

3.3   Hollow fiber system 129

3.4   Gastric retentive system 129

3.5   Sigmoidal release system 129

4.   **Applications in targeted drug delivery** **130**

4.1   Oral drug delivery 130

4.2   Nasal drug delivery 130

4.3   Transdermal drug delivery 131

4.4   Ophthalmic drug delivery 132

4.5   Application in cancer treatment 132

5.   **Applications in therapeutics** **133**

5.1   High cholesterol treatment 133

5.2   Application in treatment of pruritus 133

5.3   Applications in treating of oedema 134

5.4   Application in the treatment of cardiac oedema 135

5.5   Applications as antacids 135

5.6   Treating uremia 135

Conclusion **136**

References **136**

Ion Exchange Resins: Biomedical and Environmental Applications   Materials Research Forum LLC
Materials Research Foundations 137 (2023) 120-141   https://doi.org/10.21741/9781644902219-7

1.   Introduction

Ion exchange resin is a resin or a cross-linked polymer that can be employed for the exchange of ions in the medium owing to the presence of polymer chains containing charged salt-forming groups which can be used to exchange similarly charged counterions from the environment [1]. Ion exchange resin is fabricated from the backbone of organic polymer and normally exists in the form of whitish or yellowish colored beads of 1-2 mm in diameter. Ion exchange resins have a highly structured porous surface, which is used for the exchange of ions; a simultaneous process that involves the entrapment and release of the counter-ions from the resin's surface. They can exchange either cations or anions from the medium and based on the type of ion-exchanged from the resins these can be classified as anion exchange and cation exchange resins, replacing cations and anions respectively [2]. A common example of anionic resin is Duolite AP143 which can be used to mask the taste and odor issue associated with the drugs [3], while an example of cationic resin is Amberlite IRP69 which is used as a carrier and sustained release of the drugs [4]. The in-vitro complexation and formulation of the drug with ion exchange resin followed by in-vivo release and absorption of the drug in the body is presented in Fig. 1. The ions of the drug replace the ions on ion exchange resins, which carry the drug ions inside the body where the resins release the drug ions by replacing them from the ions inside the body. The application of ion exchange resins is however limited to only those drugs that contain charged groups, as in absence of charged groups there is no binding site for the attachment and release of the drug. Several drugs employ ion exchange resins in their formulations, some of which include ibuprofen, diclofenac, paroxetine and dextromethorphan. So, the ion exchange resins are involved in various applications associated with the drugs including table disintegration, targeted drug delivery, taste masking, imparting stability to the drug, aiding dissolution, and modified drug release [5]. Moreover, targeted delivery is another aspect of the ion exchange resins for the delivery of the drugs at the desired site and thus can be substantially employed in nasal, oral, ophthalmic, parenteral, and transdermal drug delivery. Furthermore, the resins can also be used in therapeutics where they can play a role in treating ailments or abnormalities like high cholesterol, pruritus, cardiac edema, hyperacidity, and uremia, etc. [6].

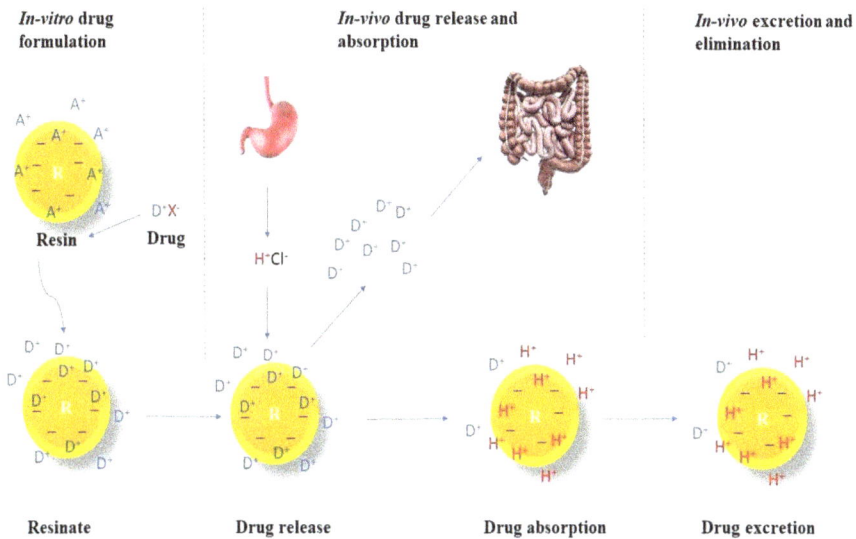

Figure 1. Drug formulation and release using ion exchange resins

2. Application of resins in formulation-related issues

2.1 Taste development

The functional group such as amine is a part of the majority of bitter drugs which is responsible for the obnoxious taste of these drugs. To reduce the bitterness of such drugs, the complex formation can be induced with some suitable agent which can result in blocking of the functional group [7-10]. Normally the drugs are converted to estolates and stearates which is usually the process used for the formation of complexes but nowadays ion exchange resins can be used for the formation of such complexes. As a result of this binding, the functional group is blocked resulting in masking the bitterness of the drugs like azithromycin and clarithromycin. The complex formation between the ion exchange resin and the drugs can be carried out either via prolonged contact of the resin with the drug solution or by repeated exposure of the drug with resin in a chromatographic column. The resins interact with the oppositely charged drugs through weak ionic bonding resulting in the formation of insoluble adsorbates or resonates. In the present time, a very important aspect for the market success of a drug in the industry is taste-making as it improves the patient's compliance thus providing recognition to the brand to combat the private-label competition [11]. Different techniques can be used to mask the undesirable taste of the drugs which include adsorption on ion exchange resin, chemical modification by using an insoluble prodrug, filling in capsules, microencapsulation with various polymers, salt formation, effervescent system, and use of excipients like surfactants, gelatin, sweeteners, gelatinized starch, and flavors. The release of the drug from the ion exchange resin depends upon two factors i.e., the properties of the resins used in ion exchange and the environment inside the gastrointestinal tract (pH and electrolyte concentration). The cationic resins which are strongly acidic in nature can play an important role in masking the bitter taste of the basic drugs. These strong acidic cationic resins can effectively function through the entire pH range. However, the weak acidic cationic resins, on the other hand, perform at a pH above 6. In the same way, strong base anion-exchange resins, although work efficiently under pH 7, are effective through the entire pH range [12]. Indion CRP-254 and Indion CRP-244 are the examples of resins that can be used to conceal the taste of bitter drugs like diphenhydramine HCl, chlorpheniramine maleate, and ephedrine [6,13].

2.2   Aiding in dissolution

The dissolution of poorly soluble drugs is slow which results in poor bioavailability of these drugs. The release rate of these drugs can be enhanced by using ion-exchange resins. The binding of the drug with the resin converts the drug to an amorphous form which induces the release of the drug leading to improved drug dissolution [14]. The drug in ion exchange resins can be released much more quickly from resonates and this release rate can be higher in comparison to the dissolution rate of the pure drugs e.g., the solubility of Indomethacin in the gastric fluid is up to ca 6 ppm however, the drug can be released more rapidly from resonates. Indomethacin, available as a commercial product, makes use of the micronization technique which involves reducing the particle size of the pharmaceutical product to achieve rapid dissolution [15]. However, the are many problems associated with with the micronization technique such as sensitivity soluble drugs due to high pressure, dust formation, agglomeration and decreasing melting point. Therefore, ion exchange resins can be an effective alternative to this [16]. The quick dissolution via ion exchange resins is due to two factors:

- The matrices used in ion exchange resins are hydrophilic which allow the water to enter the structure of resin resulting in an increase in the dissolution rate of the drugs.

- The resins also reduce the crystal lattice energy of the drugs, as the drug molecule is bound to the resin's functional site, this may further aid in the rapid dissolution of the resin-bound drug [17].

2.3   Role as disintegrating agents

The majority of the conventional agents, used as disintegrants, have rapid water uptake properties because of their swelling capacities. Ion exchange resins, although insoluble, have an affinity for water uptake due to large swelling capacities hence, act as disintegrants e.g., Indion 414 has been reported to have potential super disintegrant ability for oral drugs like montelukast sodium, roxithromycin, and dicyclomine hydrochloride [18]. As the swelling capacity of poly methacrylic carboxylic acid resin is substantial, these resins can be employed as table disintegrants e.g., Polacrilin with methacrylic acid divinylbenzene matrix [19]. Ion exchange resins used as disintegrants have several advantages over conventional disintegrating agents which are as follows:

- The water uptake capacity and the resulting swelling for the resin is comparatively rapid as compared to conventional disintegrants which results in decreasing the time for the disintegration of the drug.
- Ion exchange resins upon hydration do not show adhesive properties which prevent the lump formation resulting in imparting additional strength to the tablets.
- The ion exchange resins are effective even at low concentrations.
- The hardness of the tablets is also improved.
- The resins perform equally efficiently with both hydrophobic and hydrophilic formulations while the conventional ones are ineffective against hydrophilic formulations.

2.4   Drug stabilization

Significant research and revenue has been employed by the pharmaceutical industry in order to find the polymorphs for stabilizing the drug and increasing its solubility [17]. Ion exchange resins can deal with this problem as the amorphous products are formed as a result of the complexation of the resin and drug, which neither crystallize nor form hydrates [16]. Complexing the drug with the resinate frequently results in a product that has enhanced stability as compared to the original drug e.g., The shelf life of vitamin B12 is only a few months, however, a weak acid cationic resin can be used which binds with the vitamin B12 resulting in improved stability. While the ion exchange resins enhance the stability of vitamins, the effectiveness of the complex is also retained which is similar to that of the free form of the vitamin [20]. Another example is nicotine; it loses its color when exposed to air and light resulting in its discoloration, however, the stability of nicotine can be enhanced by complexing the nicotine with the resins as can be seen in the form of lozenges and gums of nicotine. A study has been reported regarding the stabilization of the levodopa drug in which the drug is normally oxidized which presents a problem for the storage of the drug [21]. The problem can be resolved by adjusting the pH and employing ion-exchange fibers which makes it easy to store the drug thus allowing effective drug delivery. The hygroscopic drugs, in the presence of moisture, are normally affected as they form agglomerates. The hygroscopicity of such drugs can be decreased by the adsorption of these drugs on ion exchange resins. Furthermore, the resins also provide excellent flowability to the formulations owing to their macroreticular and uniform morphology [22].

2.5   Water purification for the production of pharmaceuticals

To produce pharmaceutical products, purified water is an essential element. Cation-exchange resins can be employed for the softening of water. These resins allow the smooth running of the reverse osmosis process by preventing the formation of calcium carbonate precipitates because these resins replace the ions of calcium with sodium and the calcium ions are thus not available for forming calcium. Feedwater, in the sodium form, is trickled down the resin during which the sodium ions are released by adsorbing the polyvalent cations on the resin. The process continues till a specific time or volume after which the system is taken offline which is then automatically regenerated by using a brine solution [23].

2.6   Anti-deliquescence

The ability of a solid whereby it absorbs moisture from the surroundings to the point that it is dissolved in the adsorbed water is known as deliquescence. Deliquescence can present a problem during the production of the pharmaceutical product thus demanding the use of either specialized instruments or scheduled production during the dry season [17]. Ion exchange resins form complexes with the drug and thus can be important in addressing the above-mentioned problem for the drug. The complexation of exchange resin with highly deliquescent drugs like sodium valproate results in decreasing the uptake of moisture. The binding of the drug with the resin imparts free-flowing properties to the drug which decreases the absorbance of water. As the uptake of water decreases, the deliquescence of the drug is decreased thus aiding in stable drug production. In the same way, resins have been used to resolve similar issues for rivastigmine bitartrate [5].

3.   Applications in drug release systems

Different studies have reported the drug delivery at desired locations by using ion-exchange resins [24-31]. Some applications of ion exchange resins in drug delivery systems can be seen in Table 1. The delivery of the drugs at the desired location can have potential benefits in therapeutics as follows:

- A minimum effective concentration of the required drug is localized during the treatment.
- Systemic toxicity of the drug is reduced
- The degradation of the drug is prevented by protecting the drug from a hostile environment.

Table 1. Applications of ion exchange resins in drug delivery system

Drug	IER	Effect on Drug release	Ref.
Betahistine dihydrochloride	Dowex 50WX series	Improving the drug release	33
Chlorpheniramine	Sulfonated Styrene–Divinylbenzene Cross-linked Copolymer	Extended release of drug and improved binding	38
Cefotaxime sodium	Acrylic copolymers	Improved adsorption of the drug	36
Diclofenac	Amberlite-IRA900	Oral floating drug delivery	
Diltiazem hydrochloride	Indion 254	Ionically cross-linked microcapsules; extended release up to 15 h	35
Dextromethorphan	Dowex 50Wx2 and x4 Dowex 50Wx8	Extended delivery of two combined drugs with the equivalent therapeutic dose	45
Dextromethorphan hydrobromide	Dowex WX2-400	Aiding in sustained release with better disintegration properties	41
Diltiazem hydrochloride	Carboxyalkyl methacrylates	Release kinetics with the suitable hydrophobicity	
Domperidone	Indion 234	In-vitro and ex vivo retardation in release	37
Diphenhydramine hydrochloride	Dowex 50Wx2 and x4 Dowex 50Wx8	Improving the release of drugs in combination with equal therapeutic dose	45

Ketoprofen	poly(propylene-g-vinylbenzyl trimethylammonium-chloride)	Iontophoretically assisted transport across rat skin favored	44
Propranolol hydrochloride	Carboxyalkyl methacrylates	release kinetics with the suitable hydrophobicity	46
Potassium diclofenac	Cholestyramine	Aiding in the enteric drug delivery	43
Ranitidine hydrochloride	Indion 234	In-vitro and ex vivo retardation in release	37
Riboflavin-5-phosphate	Poly (ethylene-g-styrenetrimethyl-ammonium	Mucoadhesive properties in vivo	40
Sodium salicylate	Brominated poly(2,6-dimethyl-1,4-phenylene oxide)-methacryloxypropyl trimethoxysilane	Scaffolding material for controlled release delivery	34
Tetracycline	Carboxyalkyl methacrylates	Improved release of the drug	46
Verapamil hydrochloride	Carboxyalkyl methacrylates	Release kinetics with the suitable hydrophobicity	46

3.1   Simple resinates

One of the simplest, controlled and sustained release systems for the delivery of the drugs includes resinates which can either be suspended in a liquid, directly filled in a capsule, compressed into tablets, or suspended in matrices. Drugs in their pure form will be released and absorbed quickly from the body meanwhile the drug from the resin will be released slowly and will eventually have much better absorption. In arthritic patients, the desired rate of release for diclofenac was achieved to avoid gastric irritation [32].

3.2   Microencapsulated resinates

Microencapsulation aids in the controlled delivery of drugs at the target site as it consists of a membrane that prevents the quick release of the drug [33]. The release and absorption of the drugs from coated resins is a multi-step process starting from the entry of counter-ions in the coated resins which is followed by the drug release from the drug-resin complex and finally the absorption of the drug in the surrounding environment via diffusion of the drug ions through the membrane. Microencapsulation can be achieved by pan coating, Wurster process (air suspension coating) or by solvent evaporation [6].

3.3   Hollow fiber system

A hollow fiber system can be employed for the drug delivery purpose owing to its benefits involving loading flexibility, low gastrointestinal tract transit time, the permeability of the membrane, and high surface area to volume ratio; all these mentioned characteristics enable the efficient, controlled, effective and targeted delivery of drugs. A suitable polymeric material is used to make hollow fibers that are filled with resin in order to obtain a sustained release profile. Different studies employing in-vivo and in-vitro techniques have also been carried out for testing the sustained release of phenylpropanolamine using fiber and resin complexes [8,19].

3.4   Gastric retentive system

The drugs that are predominantly absorbed from the stomach must have a longer retention time in the gastrointestinal tract as it would increase the bioavailability of the drug to be absorbed and reduce the wastage of the drug. Some of these drugs include ciprofloxacin, furosemide, allopurinol, and cyclosporin. In order to enhance the residence of drugs inside the gastrointestinal tract, floating dosages can be an effective alternative. A study also reported the improvement of the retention time of the drug in the gastrointestinal tract by using a novel floating extended-release system [24]. Bicarbonate resin was used in the system that allowed a prolonged retention of the drug in the gastric tract as the resin consisted of a semipermeable membrane which aided in controlled release of the drug. Some of the ion exchange resins have mucoadhesive properties which majorly include anion exchange resins like cholestyramine. This might be due to the electrostatic interactions with the mucin and epithelial cell surface. Another appealing aspect involves designing specifically targeted formulations, by using bio-adhesive exchange resins, for the gastrointestinal tract. The antibiotics developed by this approach show improved localized delivery in comparison to the conventional forms of drugs e.g., tetracyclines can be delivered at the site of localized helicobacter pylori infection. Such applications for bio-adhesive ion exchange resins have been reported by different studies [28,34].

3.5   Sigmoidal release system

The concentration of the drug at the desired level can be sustained in the body by controlling the drug release at the desired location. However, this controlled release must be maintained keeping in view two important aspects i.e., the therapeutic purpose for which the drug has been designed and the properties possessed by the drug [35]. The sigmoidal release is an effective system that can help in both controlled as well as the rhythmic release of the drug at the desired target. The drug is released from the multiple unit device after a predetermined lag time which allows the rhythmic and controlled release of the drug. Ion exchange resins, as highlighted in different studies, can be used for developing the sigmoidal release systems for the sustained release of the drugs e.g., Eudragit RS, which is an anion exchange resin, maintains the ionic environment, allowing the rhythmic release of the drugs, by adding organic acid to the system [29,36].

4.   Applications in targeted drug delivery

4.1   Oral drug delivery

The use of ion-exchange resins is encouraged in the oral drug release system owing to their preferable properties which include the inert nature, sustained and equilibrium drug release, drug stability, spherical shape assisting, and retaining uniformity during the whole process [37]. A continued and prolonged release of the drug is generally considered to be preferable for the drug however, this can also cause major issues like dose dumping which enhances the risk of toxicity of the drug at the target site. The issue can be resolved by using ion exchange resins which have occupied an important place in the targeted drug delivery system as they retain the properties of the drugs while preventing the dose dumping by developing a sustained-release system. The release of the drug from the resin occurs via replacing an ion with the drug, the ion must be of the same charge as that of the drug which is to be replaced. As a similar ion has replaced the binding site of the drug therefore the homogeneity of the system is maintained. Furthermore, this exchange process is carried out at equilibrium, so the process is also dependent upon the ionic constitution, fluid volume, and body fluid composition. Moreover, the drug is not released from the resin promptly, rather a continued and prolonged release of the drug is maintained and all these phenomena are accumulatively maintained in a sustained-release system [38,39]. A study reported the preparation of microcapsules, via the emulsion solvent evaporation method, using ion exchange resins to attain the sustained release of diltiazem suspension. The drug suspension prepared by this method was stable and allowed the sustained release of the drug with low drug leaching. Furthermore, the release profiles for these suspensions also remained unchanged for 120 days at 30°C and 45°C when they were compared with the dried microcapsules. So the use of ion exchange resin provided a prepared suspension of the drug for stable oral delivery [40].

4.2   Nasal drug delivery

Amberlite, a formulation administered via nasal route, is a weakly acidic resin complex that allows the rhythmic and sustained release of nicotine from the resin with the purpose to quit the smoking. Different variants of the Amberlite have also been developed like Amberlite IRP69 and Amberlite IR120, where IRP69 provides improved adsorptive capacity and better flow property. The plasma profiles for the nicotine showed that the ratio of free to bound form of nicotine in Amberlite formulations can be adjusted so that an elevated, rhythmic and sustained concentration of nicotine following an initial plasma peak level can be achieved [41].

A study evaluated the increased immune response to influenza hemagglutination vaccine, administered through intranasal route in mice, by complexing the vaccine with ion exchange resin. The results of the study showed that the higher levels of IgA-induced mucosal immunity, hemagglutinin antibodies, and systemic immunity were achieved by complexing the resin i.e., sodium polystyrene sulfonate with the influenza hemagglutination vaccine. Furthermore, when these mice, administered with resin complexed vaccine, were compared with those having received hemagglutinin vaccine alone the results showed better protection to viral challenge in mice with the complexed vaccine. Therefore, the resin complexed influenza hemagglutination vaccine can be administered via the intranasal route possessing the potential for safe and effective immunization [42].

4.3   Transdermal drug delivery

Ion exchange resins also impart a role in formulations based on the transdermal drug delivery system. A study was carried out in which a model was postulated involving ion exchange resins with the aim to deliver the drug via a transdermal route to determine the potential for the designed model. Carbopol-based gel vehicles were used which consisted of ion exchanging fibers in order to determine the release rate of ketoprofen from the gel. For this purpose, 0.22 um microporous membrane was used across which the release rate of ketoprofen, bound to ion exchange fibers, was determined. The fluctuation in the release rate was compared for both the Carbopol-based gel vehicle and the simple gels which showed fewer fluctuation rates for Carbopol gel in comparison to simple gels. However, there was one drawback that the total concentration of the drug delivered was comparatively less; the issue could be resolved by additional ions which will increase the rate and extent of delivery of ketoprofen [43].

Another study showed the transdermal delivery of zwitterionic levodopa for controlled delivery of the drug using ion-exchange fiber and iontophoresis. A comparison was carried out in a study for the resin complexed drugs i.e., levodopa and metaraminol, with parameters like transdermal permeation of the drug. The results for the study showed better results for the cationic model metaraminol as compared to the levodopa [21].

4.4   Ophthalmic drug delivery

Ion exchange resins also play an important role in the ophthalmic delivery system. A study highlighted the effect on Betoptic S, a product involving the complexation of the antiglaucoma agent, and betaxolol hydrochloride complexed with ion exchange resins [44]. Betoptic S is a beta-adrenergic receptor blocking agent which is used to treat glaucoma and hypertension. The drug-resin combination employs the use of Amberlite1 IRP (a cation ion exchange resin) and a positively charged drug that forms a complex with the resin. A structured vehicle is also employed, containing Carbomer 934P as a polymer, for the incorporation of drug resin particles. This enhances the physical stability of the product and the product can be easily resuspended again. The prepared ophthalmic suspension (0.25%) of the drug resulted in an increase in the bioavailability of the drug. The resultant product provided a rhythmic flow of the drug providing uniform dosage and increasing the ocular comfort.

A study was performed to compare the efficacy of a designed experimental formulation with the fluoroquinolone by employing an experimental rabbit model of Staphylococcus keratitis [12]. The experimental formulation consisted of a combination of drug and resin i.e., ciprofloxacin-PSS (ciprofloxacin complexed with polystyrene sulfonate). So, ion exchange resin was used for the delivery of the drug to the cornea. The study showed effective topical treatment of Staphylococcal keratitis with ciprofloxacin and ciprofloxacin-PSS. The results for the ciprofloxacin-PSS suggested that the ion exchange resins can be employed in ocular fluoroquinolone formulation for the improved drug delivery system.

4.5   Application in cancer treatment

The anticancer drugs can be entrapped inside the carrier molecules, like microcapsules and microspheres, which is a commonly used approach for the delivery of the anticancer drugs at the target site. The anticancer drugs can be delivered at the target site by using ion exchange resins because these drugs can bind with the resin owing to the ionic nature of the anticancer drugs. Different studies have reported the use of ion exchange resins for the site-specific delivery of anticancer drugs to treat the cancerous cells [6,37]. The mechanism for the delivery of doxorubicin complexed with ion-exchange resin has also been studied for anticancer therapy. The studies showed that ion exchange resins can be employed for drug loading and delivery at their maximum level at the targeted site [35].

Materials Research Forum LLC
https://doi.org/10.21741/9781644902219-7

5.   Applications in therapeutics

Ion exchange resins also find value in therapy, with prominent roles in the treatment of several diseases. Renal disease at the end-stage can be dealt with using cationic resins in oral therapeutics by reducing the phosphate levels meanwhile hyperkalemia associated with acute renal can be treated by using cation exchange resins such as sodium polystyrene sulphonate which is employed as an adjuvant in the treatment of the disorder [45]. So, ion exchange resins can be employed in different pathological states like treatment of cholesterol, pruritus, cardiac edema, hyperacidity, nephrotic, sodium and potassium supplement depletion, and ulcers, etc.

5.1   High cholesterol treatment

The first approved resin-based drug for the treatment of high cholesterol is cholestyramine. It is a cationic resin that enters the gastrointestinal tract and binds with bile acids resulting in the sequestration of bile [46]. As the bile acid is bound and removed, the deficiency of the bile acid signals the liver to synthesize it for which the cholesterol is consumed, as it is required for the bile acid production, and as a result the cholesterol in the body is reduced. The competitive advantage for the above-mentioned treatment is that it has only a few side effects as it does not make use of the conventional drugs with different after-effects. However, the therapy has a disadvantage as higher doses of the resins are required i.e., four tablespoons administered through fruit juice. This issue was faced mostly with the first-generation resins but with the subsequent upcoming models, the dose level of the resins has decreased from four tablespoons to only a regular pharmaceutical-sized capsule. This was done by employing molecular modeling strategies in order to enhance the specificity of the resins for the bile acid which in turn reduced the concentration of resin to be administered. The specificity for the resins was enhanced by using polymers which improved the binding properties of the resin by employing multiple binding interactions i.e., electrostatic, hydrophobic. [38].

Different modified and improved bile acid sequestering resins have also been developed which are more effective than the conventional resin e.g., Colestimide, a next-generation bile acid sequestering resin, is a polymer of 2-methylimidazole-epichlorohydrin which is four times powerful in comparison to the cholestyramine. Colestimide is also effective in Type-2 Diabetes patients with hypercholesterolemia which has also been reported to reduce blood glucose levels [47].

5.2   Application in treatment of pruritus

Non-adsorbable ion exchange resins are not absorbed from the gastrointestinal tract and inside the lumen, these resins release chloride and bind to bile acids resulting in their decreased recirculation. So, these ion exchange resins enhance the excretion of the cholesterol by conversion to bile acids. Bile acid in turn dissociates quickly and is absorbed in the ileum. Cholestyramine is an orange-flavored granule that can be employed for improving pruritus [48]. Cholestyramine has been used for a long time owing to its proven safety and effectiveness, however, it has poor palatability which limits the tolerance of the drug. The issue can be resolved by administering the drug via a nasogastric tube which improves the tolerability and intake of the drug resulting in better control of cholesterol levels and improving pruritus. Colesevelam is another drug that is employed as tablets or non-adsorbable hydrogels; it is probably the best-tolerated drug having preferable properties such as enhanced specificity, better affinity, and better ability to bind bile acids in comparison to several other non-adsorbable ion exchange resins. However, an issue with the use of non-adsorbable ion exchange resins is that they pose problems for the absorption of some of the drugs such as ursodeoxycholic acid. Therefore these drugs should be administered almost 4-6 h after the administration of cholestyramine. This would allow better functioning of the non-adsorbable ion exchange resins while preventing interaction with other drugs. The normal dosage of these resins is 8-16 g which can be administered orally 2-3 times daily while progressively increasing by 2g daily. The use of non-adsorbable ion exchange resins can cause the release and adsorption of a large quantity of chloride in the gastrointestinal tract which results in constipation and hyperchloremic acidosis [6].

5.3   Applications in treating of oedema

Cation-exchange resins have a variable affinity for different ions i.e., greater affinity with magnesium and calcium as compared to potassium while better affinity for potassium in comparison to sodium. In a solution, the resins can replace one cation with another under favorable conditions. The extent to which the ion exchange can occur depends upon the pH of the medium, the relative concentration of ions, and the time allowed for the process to take place. Resins can be employed for the treatment of edema such as carboxylic or sulphonic resins, which requires the administration of these resins, in hydrogen or ammonium form, via oral route. The upper part of the gastrointestinal tract is the ion exchanging site where the major exchange of cations take place. The lost cations are replaced mainly through the cations ingested from the food. Bicarbonates can be used to fulfill the loss of anions from the extracellular fluid. Furthermore, a large quantity of ammonia is produced after a few days of delay which aids in further accommodation of the loss. So, the sodium ions are excreted outside the body through feces and to replace the sodium ions, the disposable cations are absorbed. Kidney excretes the corresponding anions from the body, which is done by allowing a minimum change in the pH of the plasma. The concentration of sodium released from the body depends on the sodium intake, as the dietary intake decreases the removal of sodium via feces is also decreased, and it is rare to have more sodium in the fecal matter than that of the diet. [6].

5.4   Application in the treatment of cardiac oedema

Cardiac edema is a congestive heart failure that can be treated by using a known ammonium-potassium carboxylic resin i.e., Resodec, having efficacy comparable to that of dietetic sodium restriction. However, side effects can sometimes be observed due to the abnormal potassium level in serum or due to hyperchloremia. For the patients with sodium-restricted diet or having a renal impairment or those on mercurial diuretics, it is essential to have biochemical control at the early stages of the treatment as the previous treatments may result in abnormal blood electrolytes levels in these patients. The use of ion-exchange resins in abnormalities results in its treatment but sometimes there is a possibility of aggravating the situation. To obtain the best results through the resin treatment, ammonium chloride must not be administered during resin uptake and the treatment must be carried out with repeated short courses [6].

5.5   Applications as antacids

Anion exchange resins can be employed for both ulcer and non-ulcer patients where they play an important role in inactivating pepsin and neutralizing acidity [49]. The claimed benefits for the anion exchange resins include:

- Working action of the resin is at great speed
- Effective neutralizing power when employed in practical applications
- Comprehensive peptic activity inhibition
- No chloride removal
- Lack of acid rebound
- No sign of side effects like constipation

5.6   Treating uremia

The working properties of the weakly acidic drugs, such as the serum binding ability, are diminished in case of abnormalities like uremia. This issue can be addressed by employing anion exchange resins; Amberlite CG-400 (anion exchange resin) has been used for the treatment of uremic serum where it resolved the binding defect of three drugs i.e., salicylate, nafcillin, and sulfamethoxazole [50]. The binding of the basic drugs with serum protein however remains unaffected by uremia and the induction of resins doesn't modify the attachment of these drugs, as was observed in the case of the above-mentioned example where the Amberlite CG-40 did not affect drug attachment to its target site i.e., quinidine and trimethoprim. The defects in the binding capacity in uremia are based on different factors which appear to be lipid solubility, dialyzable, and weakly acidic. Unpredictable results were obtained for the effects of resin treatment on the binding of penicillin G which were similar to the one with the activated charcoal treatment [51] and butyl chloride extraction [52].

Another study demonstrated the positive relationship between the degree of penicillin G binding and the concentration of free fatty acids determining the influence of free fatty acids on the ability of penicillin G to bind with proteins [13]. Anion exchange resins can be employed for treating the serum specimens which can result in reducing the binding of penicillin G by removing free fatty acids. In cases of unconjugated hyperbilirubinemia, which results in decreased conjugation of bilirubin often resulting in jaundice in infants, a hemoperfusion system can be developed consisting of the ion exchange resin particles in order to remove bilirubin which can aid in the treatment of jaundiced infants. A macro reticular resin in the form of a packed bed is used to develop a hemoperfusion system, which is coated by an albumin monomolecular layer to make it biocompatible. Different factors dictate increased adsorption of bilirubin and better blood compatibility of the resin which include choosing the right ionic resin, the proper coating of albumin and the procedure of cross-linking. The albumin-coated resin is highly effective and can efficiently remove 80-90% of bilirubin, in-vitro, from the plasma [6].

Conclusion

Ion exchange resins can be employed for various applications in pharmaceutics apart from their commonly known roles of separation, purification, and processing. For decades, ion exchange resins have been employed for traditional purposes; however, various studies have highlighted the potential of ion exchange resins in the pharmaceutical industry for sustained drug release, better disintegration, aiding in dissolution, and taste masking for the drugs. So, the ion exchange system has the potential to improve the stability, efficacy, and safety of the drug, therefore, increasing the patient's compliance. Furthermore, the therapeutic applications may open up new avenues and this may help researchers in developing better drug delivery and treatment options

References

[1] S.D. Alexandratos. Ion-exchange resins: a retrospective from industrial and engineering chemistry research. Industrial & Engineering Chemistry Research, 48(1) (2009) pp.388-398. https://doi.org/10.1021/ie801242v

[2] M.A. Harmer and Q. Sun. Solid acid catalysis using ion-exchange resins. Applied Catalysis A: General, 221(1-2) (2001) 45-62. https://doi.org/10.1016/S0926-860X(01)00794-3

[3] S. Sivaneswari, E. Karthikeyan, D. Veena, P.J. Chandana, P. Subhashree, L. Ramya, R. Rajalakshmi, A.K. CK, Physicochemical characterization of taste masking levetiracetam ion exchange resinates in the solid state and formulation of stable liquid suspension for pediatric use, Beni-Suef Univ. J. Basic Appl. Sci. 5 (2) (2016) 126-133. https://doi.org/10.1016/j.bjbas.2016.04.004

[4] S.H. Jeong, K. Park, Development of sustained release fast-disintegrating tablets using various polymer-coated ion-exchange resin complexes, Int. J. Pharm. 353 (1-2) (2008) 195-204. https://doi.org/10.1016/j.ijpharm.2007.11.033

[5] L. Hughes, Ion exchange resins unique solutions to formulation problems, Pharm. Technol. 22 (2004) 20-25.

[6] S.N. Khan, Therapeutic applications of ion exchange resins, in: Inamuddin, M. Luqman (Eds.), Ion Exchange Technology II, Springer, Dordrech,t 2012, pp. 149-168. https://doi.org/10.1007/978-94-007-4026-6_7

[7] W.P. Arnold Jr., Medical uses of ion-exchange resins, N. Engl. J. Med. 245 (9) (1951) 331-336. https://doi.org/10.1056/NEJM195108302450905

[8] M.A. Hussain, R.C. DiLuccio, E. Shefter, Hollow fibers as an oral sustained-release delivery system, Pharm. Res. 6(1) (1989) 49-52. https://doi.org/10.1023/A:1015847618671

[9] S. Pande, M. Kshirsagar, A. Chandewar, Ion exchange resins pharmaceutical applications and recent advancement, Int. J. Adv. Pharm. 2 (1) (2011).

[10] L.S. Goodman. Goodman and Gilman's the pharmacological basis of therapeutics, New York: McGraw-Hill. Vol. 1549, (1990) pp. 1361-1373.

[11] N.C. Chaudhry, L. Saunders, Sustained release of drugs from ion exchange resins, J. Pharm. Pharmacol. 8 (1) (1956) 975-986. https://doi.org/10.1111/j.2042-7158.1956.tb12227.x

[12] J.M. Moreau, L.C. Green, L.S. Engel, J.M. Hill, R.J. O'Callaghan, Effectiveness of ciprofloxacin-polystyrene sulfonate (PSS), ciprofloxacin and ofloxacin in a Staphylococcus keratitis model, Curr. Eye Res. 17 (8) (1998) 808-812. https://doi.org/10.1080/02713689808951262

[13] B. Suh, W.A. Craig, A.C. England, R.L. Elliott, Effect of free fatty acids on protein binding of antimicrobial agents, J. Infect. Dis. 143 (4) (1981) 609-616. https://doi.org/10.1093/infdis/143.4.609

[14] D. Shukla, S. Chakraborty, S. Singh and B. Mishra. Mouth dissolving tablets II: An overview of evaluation techniques. Scientia Pharmaceutica, 77(2) (2009) pp.327-342. https://doi.org/10.3797/scipharm.0811-09-02

[15] V. Suhagiya, A. Goyani, R. Gupta, Taste masking by ion exchange resin and its new applications: A review, Int. J. Pharm. Sci. Res. 1 (4) (2010) 22-37.

[16] S. Vijay, C. Dr, Ion exchange resins and their applications, J. Drug Deliv. Ther. 4 (4) (2014) 115-123. https://doi.org/10.22270/jddt.v4i4.925

[17] L. Hughes, New uses of ion exchange resins in pharmaceutical formulation, Pharm. Tech. 17 (2005) 38.

[18] P. Amin, N. Prabhu, and A. Wadhwani. Indion 414 as superdisintegrant in formulation of mouth dissolve tablets. Indian journal of pharmaceutical sciences, 68(1) (2006) p.117. https://doi.org/10.4103/0250-474X.22983

[19] J. Mahore, K. Wadher, M. Umekar, P. Bhoyar, Ion exchange resins: pharmaceutical applications and recent advancement, Int. J. Pharm. Sci. Res. 1 (2) (2010) 8-13.

[20] K. Hänninen, A.M. Kaukonen, T. Kankkunen, J. Hirvonen, Rate and extent of ion-exchange process: the effect of physico-chemical characteristics of salicylate anions, J. Control Release. 91 (3) (2003) 449-463. https://doi.org/10.1016/S0168-3659(03)00276-1

[21] T. Kankkunen, I. Huupponen, K. Lahtinen, M. Sundell, K. Ekman, K. Kontturi, J. Hirvonen, Improved stability and release control of levodopa and metaraminol using ion-exchange fibers and transdermal iontophoresis, Eur. J. Pharm. Sci. 16 (4-5) (2002) 273-280. https://doi.org/10.1016/S0928-0987(02)00113-6

[22] M. Chaubal, Synthetic polymer-based ion exchange resins: excipients & actives, Drug Deliv. Tech. 3 (5) (2003).

[23] A. Bennett, High purity water: Advances in ion exchange technology, Filtr. Sep. 44 (6) (2007) 20-23. https://doi.org/10.1016/S0015-1882(07)70180-5

[24] J.T. Fell, L. Whitehead, J.H. Collett, Prolonged gastric retention: using floating dosage forms, Pharm. Technol. 24 (3) (2000) 82-90.

[25] Y. Chen, M.A. Burton, J.P. Codde, S. Napoli, I.J. Martins, B.N. Gray, Evaluation of Ion-exchange Microspheres as Carriers for the Anticancer Drug Doxorubicin: In-vitro

Studies, J. Pharm. Pharmacol. 44 (3) (1992) 211-215. https://doi.org/10.1111/j.2042-7158.1992.tb03583.x

[26] S. Burton, N. Washington, R. Steele, R.M. FEELY, Intragastric distribution of ion-exchange resins: A drug delivery system for the topical treatment of the gastric mucosa, J. Pharm. Pharmacol. 47 (11) (1995) 901-906. https://doi.org/10.1111/j.2042-7158.1995.tb03268.x

[27] F. Atyabi, H. Sharma, H. Mohammad, J. Fell, Controlled drug release from coated floating ion exchange resin beads, J. Control Release. 42 (1) (1996) 25-28. https://doi.org/10.1016/0168-3659(96)01343-0

[28] W. Irwin, R. Machale, P. Watts, Drug-delivery by ion-exchange. Part VII: release of acidic drugs from anionic exchange resinate complexes, Drug Dev. Ind. Pharm. 16 (6) (1990) 883-898. https://doi.org/10.3109/03639049009114916

[29] S. Narisawa, M. Nagata, Y. Hirakawa, M. Kobayashi, H. Yoshino, An organic acid-induced sigmoidal release system for oral controlled-release preparations. III. Elucidation of the anomalous drug release behavior through osmotic pumping mechanism, Int. J. Pharm. 148 (1) (1997) 85-91. https://doi.org/10.1016/S0378-5173(96)04834-X

[30] S. Jackson, D. Bush, A. Perkins, Comparative scintigraphic assessment of the intragastric distribution and residence of cholestyramine, Carbopol 934P and sucralfate, Int. J. Pharm. 212 (1) (2001) 55-62. https://doi.org/10.1016/S0378-5173(00)00600-1

[31] F. Siepmann, J. Siepmann, M. Walther, R. MacRae, R. Bodmeier, Polymer blends for controlled release coatings, J. Control Release. 125 (1) (2008) 1-15. https://doi.org/10.1016/j.jconrel.2007.09.012

[32] M. Kurowski, H. Menninger, E. Pauli, The efficacy and relative bioavailability of diclofenac resinate in rheumatoid arthritis patients, Int. J. Clin. Pharmacol. Ther. 32 (8) (1994) 433-440.

[33] A.U. Kadam, D.M. Sakarkar and P.S. Kawtikwar. Development and evaluation of oral controlled release chlorpheniramine-ion exchange resinate suspension. Indian Journal of Pharmaceutical Sciences, 70(4) (2008) p.531. https://doi.org/10.4103/0250-474X.44613

[34] S. Burton, N. Washington, R. Steele, R.M. FEELY, Intragastric distribution of ion-exchange resins: A drug delivery system for the topical treatment of the gastric

mucosa, J. Pharm. Pharmacol. 47 (11) (1995) 901-906. https://doi.org/10.1111/j.2042-7158.1995.tb03268.x

[35] A. Sawaya, J.P. Benoit, S. Benita, Binding mechanism of doxorubicin in Ion-exchange albumin microcapsules, J. Pharm. Sci. 76 (6) (1987) 475-480. https://doi.org/10.1002/jps.2600760613

[36] R.K. Kamble, C.C. Singh, K.R. Priyadarshani.Current Trends: Ion Exchange Resinates in Controlled Drug Delivery. International Journal of Pharmaceutics and Drug Analysis, 2(1) (2014) 1-11.

[37] V. Anand, R. Kandarapu, S. Garg, Ion-exchange resins: carrying drug delivery forward, Drug Discov. 6 (17) (2001) 905-914. https://doi.org/10.1016/S1359-6446(01)01922-5

[38] I. Singh, A.K. Rehni, R. Kalra, G. Joshi, M. Kumar, H.Y. Aboul-Enein, Ion exchange resins: Drug delivery and therapeutic applications, Fabad J. Pharm. Sci. 32 (2) (2007) 91.

[39] L. Hughes, Ion Exchange Resinates, Pharm. Technol. Eur. 17 (4) (2005) 38-42.

[40] V.B. Junyaprasert, G. Manwiwattanakul, Release profile comparison and stability of diltiazem-resin microcapsules in sustained release suspensions, Int. J. Pharm. 352 (1-2) (2008) 81-91. https://doi.org/10.1016/j.ijpharm.2007.10.018

[41] Y.-H. Cheng, P. Watts, M. Hinchcliffe, R. Hotchkiss, R. Nankervis, N. Faraj, A. Smith, S. Davis, L. Illum, Development of a novel nasal nicotine formulation comprising an optimal pulsatile and sustained plasma nicotine profile for smoking cessation, J. Control Release. 79 (1-3) (2002) 243-254. https://doi.org/10.1016/S0168-3659(01)00553-3

[42] M. Higaki, T. Takase, R. Igarashi, Y. Suzuki, C. Aizawa, Y. Mizushima, Enhancement of immune response to intranasal influenza HA vaccine by microparticle resin, Vaccine. 16 (7) (1998) 741-745. https://doi.org/10.1016/S0264-410X(97)00248-X

[43] L. Yu, S. Li, Y. Yuan, Y. Dai, H. Liu, The delivery of ketoprofen from a system containing ion-exchange fibers, Int. J. Pharm. 319 (1-2) (2006) 107-113. https://doi.org/10.1016/j.ijpharm.2006.03.039

[44] R. Jani, O. Gan, Y. Ali, R. Rodstrom, S. Hancock, Ion exchange resins for ophthalmic delivery, J. Ocul. Pharmacol. Ther. 10 (1) (1994) 57-67. https://doi.org/10.1089/jop.1994.10.57

[45] M.V. Srikanth, S.A. Sunil, N.S. Rao, , M.U. Uhumwangho and K.R. Murthy. Ion-exchange resins as controlled drug delivery carriers. Journal of Scientific Research, 2(3) (2010) pp.597-597. https://doi.org/10.3329/jsr.v2i3.4991

[46] A.J. Hilmer, R.B. Jeffrey, W.G. Park, C. Khosla, Cholestyramine as a promising, strong anion exchange resin for direct capture of genetic biomarkers from raw pancreatic fluids, Biotechnol. Bioeng. 114 (4) (2017) 934-938. https://doi.org/10.1002/bit.26207

[47] T. Suzuki, K. Oba, Y. Igari, N. Matsumura, K. Watanabe, S. Futami-Suda, H. Yasuoka, M. Ouchi, K. Suzuki, Y. Kigawa, Colestimide lowers plasma glucose levels and increases plasma glucagon-like PEPTIDE-1 (7-36) levels in patients with type 2 diabetes mellitus complicated by hypercholesterolemia, J. Nippon. Med. Sch. 74 (5) (2007) 338-343. https://doi.org/10.1272/jnms.74.338

[48] F. Alvarez, Treatments in chronic cholestasis in children, Ann Nestlé [Engl]. 66 (3) (2008) 127-135. https://doi.org/10.1159/000147410

[49] M. Srikanth, S. Sunil, N. Rao, M. Uhumwangho, K.R. Murthy, Ion-exchange resins as controlled drug delivery carriers, J. Sci. Res. 2 (3) (2010) 597-597. https://doi.org/10.3329/jsr.v2i3.4991

[50] D.M. Lichtenwalner, B. Suh, B. Lorber, M.R. Rudnick, W.A. Craig, Correction of drug binding defects in uremia in vitro by anion exchange resin treatment, Biochem. Pharmacol. 31 (21) (1982) 3483-3487. https://doi.org/10.1016/0006-2952(82)90630-X

[51] W.A. Craig, M.A. Evenson, K.P. Sarver, J.P. Wagnild, Correction of protein binding defect in uremic sera by charcoal treatment, J. Lab. Clin. Med. 87 (4) (1976) 637-647.

[52] D. Lichtenwalner, B. Suh, B. Lorber, M. Rudnick, W. Craig, Partial purification and characterization of the drug-binding-defect inducer in uremia, J. Lab. Clin. Med. 97 (1) (1981) 72-81.

Ion Exchange Resins: Biomedical and Environmental Applications   Materials Research Forum LLC
Materials Research Foundations 137 (2023) 142-165   https://doi.org/10.21741/9781644902219-8

Chapter 8

Applications of Ion Exchange Resins in Water Softening

Yu. Dzyazko

V.I. Vernadskii Institute of General and Inorganic Chemistry of the National Academy of Science of Ukraine

dzyazko@gmail.com

Abstract

Surface and groundwater always contain hardness ions (Ca^{2+} and Mg^{2+}). The hardness is an important characteristic that provides the consumer properties of water. This parameter must be taken into consideration by the stations of water treatment, thermal power plants, the enterprises of chemical, food, pharmaceutical industries. Ion exchange resins, which are intended for water softening, are considered in this chapter. The negative effect of hardness ions on human health and equipment is also a focus of attention. Special approaches for increasing the efficiency of water softening are also reported. These approaches involve combining ion exchange with electrodialysis or ultrasound.

Keywords

Water Softening, Hardness Ions, Ion Exchange Resins, Polymer-Inorganic Resins, Electrodeionization

Contents

Applications of Ion Exchange Resins in Water Softening 142

1.   Introduction 143

2.   Water hardness 144

   2.1   Salts providing hardness 144

   2.2   Negative effect of water hardness 146

3.   Ion exchange resins for water softening 148

3.1   Strongly acidic resins 149

3.2   Weakly acidic resins 150

3.3   Polymer-inorganic resins 151

4.   Regeneration of ion exchange resins and their fouling 154

5.   Ion exchange in a combination with other processes 155

5.1   Ion exchange and ultrasound 155

5.2   Ion exchange and electrodialysis 156

Conclusions 157

References 158

1.   Introduction

Surface and groundwater always contain hardness ions (Ca^{2+} and Mg^{2+}), their content is various in different water sources, which are in one or other region. This parameter is an important characteristic that provides consumer properties of water. The hardness must be taken into consideration by the plants, which use a huge amount of water, namely the station of water treatment for household needs, thermal power plants, the enterprises of chemical, food, pharmaceutical industries and so on. The hardness of water determines the possibility to use it for drinking, since the increased content of Ca^{2+} and Mg^{2+} ions adversely affects human health.

For decreasing the content of hardness ions, different softening techniques have been developed. It is necessary to mention baromembrane methods: combination of chemical precipitation with microfiltration [1], polymer- assisted ultrafiltration [2] (as shown in the example of whey filtration, ultrafiltration membrane rejects up to 10-15% of hardness ions [3] without addition of macromolecular compound to the liquid being desalinated), nanofiltration [4], reverse osmosis [5]. Chemical methods (precipitation of insoluble calcium and magnesium compounds with caustic [6] and lime [7] soda) can be related to very old times and are widespread. Distillation has been widely used for a very long time [8]. Modern direction in the field of water softening is membrane distillation [9]. A number of electrochemical methods are also reported: electrodialysis [10], capacitive deionization [11], membrane capacitive deionization [12], electro precipitation [13], electrocoagulation [14].

The most common technique for water softening is ion exchange using polymer resins, since it is relatively cheap, effective and requires no energy consumptions, complex equipment and expensive materials [15]. The task of the specialists, who occupy the field of material engineering, is to develop ion exchange resins with high capacity, selectivity and fast sorption. Moreover, the resins must be easily regenerated and must be stable against regenerating solutions. Large manufacturing companies (Dow chemical, Bayer etc.) offer a wide range of ion exchange resins, which are intended for one or several purposes.

In this chapter, ion exchange resins (particularly polymer-inorganic composites), which are intended for this purpose, are considered. Negative effect of hardness ions on human health and equipment is also a focus of attention. Special approaches, which allow us to increase the efficiency of water softening are also reported . These approaches involve combining ion exchange with electrodialysis (electrodeionization) or ultrasound.

2.   Water hardness

2.1   Salts providing hardness

The permanent hardness is due to multivalent inorganic cations (i.e. cations, a charge of which is higher than +1) [1]. These cations are able to form insoluble compounds due to chemical interaction with different substances, such as soda, alkali and so on. Among multivalent cations, namely Ca^{2+} and Mg^{2+} are found in large quantities in most natural waters. At the same time, Ba^{2+}, Sr^{2+}, Fe^{3+} and multivalent cations of other metals provide water hardness, however, their contribution is inconsiderable. Thus, only calcium and magnesium cations are related to hardness ions.

Ca^{2+} and Mg^{2+} ions appear in surface and groundwater (rainwater and distilled water contain a small amount of them) due to a leakage from calcium and magnesium-containing minerals, among which calcite $CaCO_3$, anhydrite $CaSO_4$, alabaster $CaSO_4$ $0.5H_2O$, and gypsum $CaSO_4 \cdot 2H_2O$, dolomite $MgCO_3 \cdot CaCO_3$ and magnesite $MgCO_3$ are the most widespread. For instance, dissolution of calcium carbonate (bicarbonate formation) occurs as:

$$CaCO_3 \text{ (s)} + H_2O + CO_2 \text{ (aq)} \leftarrow\rightarrow Ca^{2+} \text{ (aq)} + 2\ HCO^- \text{ (aq)}\quad (1)$$

Rain water contains dissolved CO_2 gas, which interacts with calcium carbonate. As a result, a soluble calcium compound is formed. However, this reaction is reversible. The Ca carbonate can be re-deposited, its morphology strongly depends on the CO_2 content in air stalactites and stalagmites can be formed under certain conditions. The biota effect on this process is considerable, since it affects the amount of this gas in air [17].

For the numerical expression of water hardness, the concentration of Ca^{2+} and Mg^{2+} ions is indicated. The SI unit for measuring concentration is mol m^{-3}, however, in practice, degrees of hardness or mg l^{-1}, mmol l^{-1}, mg-eq l^{-1}are used to measure hardness. In different countries, various non-systemic units are used, they are so-called hardness degrees, which reflect mainly the total content of hardness ions (Table 1).

Table 1. Units of hardness ions

Country	Symbol	Definition	Value (mmol g^{-1})
Germany	dH (Deutsche Härte); dGH (degrees of general hardness); °dKH (for temporary hardness).	1 part of CaO or 0.719 parts of MgO per 100 000 parts of water	0,1783
UK	°e	1 grain CaCO$_3$ per 1 english gallon of water	0.1424
France	°TH	1 part of CaCO$_3$ per 100 000 parts of water	0.0999
USA	ppm	1 part of CaCO$_3$ per 1 000 000 parts of water	0.0100
	gpg (grain per gallon)	1 part of CaCO$_3$ per 1 american gallon of water	0.1710

Total water hardness includes (i) temporary and (ii) permanent hardness (Fig. 1). Temporary hardness is due to soluble calcium bicarbonate and magnesium bicarbonate, which are formed according to scheme (1). This hardness can be removed by boiling, in this case the reaction (1) occurs in an opposite direction. Permanent hardness, which is caused by sulfate and chloride salts, can not be removed by this manner. In this case, the approaches, which are listed above, are used. In general, they are applied to remove the total hardness.

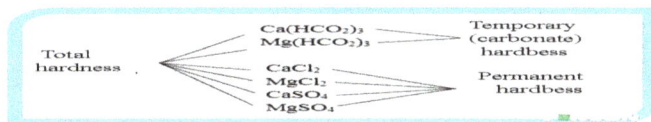

Figure 1. Types of water hardness.

In order to determine water hardness, the standard technique is used. The method is based on titration with ethylenediaminetetraacetic acid (EDTA) [18]. Analytical techniques, such as atomic absorption spectrometry (AAS) and induced coupled plasma analysis (ICP) can be also applied. Potentiometric sensors for monitoring water hardness have been proposed (the sensor also determines the presence of other ions) [19]. Different types of sensors are considered also in [20-22]. The method of quantitative analysis of hardness ions has been invented: it is based on the change of color of silver nanoparticles [23].

According to the amount of Ca^{2+} and Mg^{2+}ions, following water classification has been proposed (Table 2) [18]. Normative content of these ions for drinking water is 25-130 (Ca^{2+}) and 5-65 (Mg^{2+}) mg l^{-1}. Since hardness ions are important for living organisms (calcium is a constituent of bone tissues, magnesium is important for the normal functioning of the cardiovascular, nervous systems, digestive organs, muscles and bones), water with lower content of these species is injurious to health. At the same time, their higher concentration provides negative effects. This problem will be considered further.

Table 2. Types of water according to the content of hardness ions (reproduced according to the permission of Cambridge University Press)

Type of water	Total hardness (mg l^{-1} of $CaCO_3$)
Soft	0-50
Moderately soft	50-100
Slightly hard	100-150
Moderately hard	150-250
Hard	250-350
Very hard	>350

2.2   Negative effect of water hardness

The content of water in human organisms is 70-80%. Namely water provides transport of oxygen, enzymes, hormones, and salts in the body. The chemical composition of water becomes especially important: the more impurities it contains, the worse it dissolves useful substances. As mentioned above, the content of hardness ions in drinking water must be in a strictly determined diapason.

As suggested by the World Health Organization, no fatal health effect of hard water on human health is known [25]. Moreover, hard water provides a contribution to Ca and Mg, since deficiency of which is responsible particularly for cardiovascular and bone diseases and even cancer [26]. Drinking water is more important as a source of calcium and magnesium than diet, since water contains the most digestible compounds of these metals. Nevertheless, soft water is necessary to prepare food beverage productions, such as juices, beer, wines and so on [27]. Hard water sufficiently deteriorates the quality of products. It can be harmful in the case of medical application, for instance, hemodialysis [28] and medicine preparation [29]. In the first case, hardness ions penetrate from physiological solution through the membrane into blood changing its salt composition. In the second case, hardness ions can interact with active components of pharmaceutical drugs. Moreover, they also appear in blood together with drugs for injections. Hard water is undesirable for daily skin and hair care.

However, hardness ions in excess quantities affect human health negatively. First of all, high hardness worsens the organoleptic properties of drinking water, giving it a bitter taste. Ca^{2+} cations interact with phosphate ions directly in digestive organs decreasing the intake of phosphorus, which is also necessary for living organisms [25]. The imbalance between calcium and phosphorus causes hypercalcemia [30]. This disease is characterized by an increase in calcium concentration in blood more than 2.6 mmol l^{-1}. As a result, the excessive production of parathyroid hormone occurs. This causes a deficiency of the element in the bone tissue. This pathology can be hereditary and can develop against the background of: growth of malignant tumors; adenomas; benign tumor of the thyroid gland; overdose of vitamins A and D; diseases of the blood and kidneys; taking diuretics.

An excess of Mg causes hypermagnesemia (an increase in the concentration of magnesium in the blood plasma above 1.1 mmol l^{-1} M or 2.9 mmol l^{-1} in severe form). Symptoms include weakness, disorientation, depressed breathing, and decreased reflexes. Hypotension and heart failure are also possible. Increased content of magnesium in water causes diarrhea. Alternatively, water containing high amounts of $MgSO_4$ causes a laxative effect.

No correlation between water hardness and formation of calcium stones in kidneys was found [25]. However, some studies assume that the usage of soft water for drinking prevents calcium nephrolithiasis and the formation of kidney stones [31]. The negative effect of calcium excess on the reproductive system has been shown [32, 33].

At last, the negative effect of hardness ions on equipment (water pipes, heaters, heat exchangers, pumps etc.) due to corrosive effect and formation of insoluble compounds is well known (Fig. 2). Both corrosion and deposition of insoluble compounds damage equipment. This is another reason for the softening of water before its usage. The ion-exchangers, which are widely used for this purpose, are considered further.

a

b

Figure 2. Deposits inside the water tube and heater, which are caused by hardness ions.

3. Ion exchange resins for water softening

3.1 Strongly acidic resins

Strongly acidic resins, such as Dowex 50X8, Dowex HCR-S (Dow Chemical), Purolite C 100E (Purolyte International), Lewatit-C249 (Bayer), are often used for water softening. They are gel-like styrene-divinylbenzene polymers containing $-SO_3$ groups (Fig. 3a). As a rule, the content of cross-linking agent (divinylbenzene) in resins for this purpose is 8-10%. According to the information of producing companies, water content in H-forms of swollen resins is about 50 %, total capacity towards Na+ ions is 1.8-1.9 mmol cm^{-3} for singly charged ions. The capacity is independent of solution acidity at pH>2 (potentiometric curve is given in Fig. 4 [34]). The capacity per 1 g of dry resin is 4.8 mmol g^{-1}.

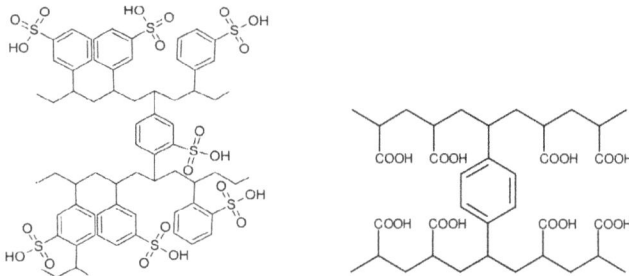

Figure 3. Polymers, which are a base of (a) strongly and (b) weakly acidic resins.

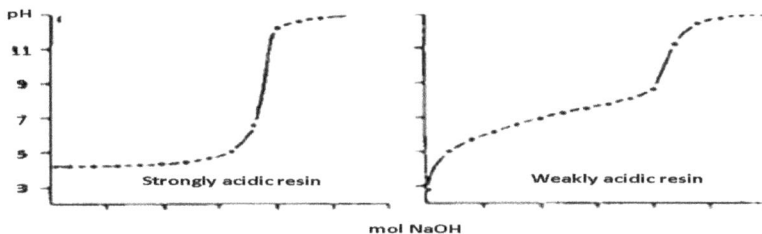

Figure 4. Titration curves of cation exchange resins [34]. Reproduced according to the permission of Elsevier Publisher.

Usually Na-forms of resins are used for water softening. Ion exchange occurs according to the scheme:

$$2R{-}SO_3^- \, Na^+ \, (s) + Ca^{2+}(aq) \rightarrow (R{-}SO_3^-)_2Ca^{2+} \, (s) + 2 \, Na^+(aq) \qquad (2)$$

$$2R{-}SO_3^- \, Na^+ \, (s) + Mg^{2+}(aq) \rightarrow (R{-}SO_3^-)_2Mg^{2+} \, (s) + 2 \, Na^+(aq) \qquad (3)$$

Here R is styrene-divinylbenzene matrix. The usage of Na-loaded forms allows one to avoid acidification of water due to releasing H^+ ions ($Ca^{2+}{\rightarrow}H^+$ and $Mg^{2+}{\rightarrow}H^+$ exchange).

Isotherm of calcium sorption were obtained for the *Purolite C100E* cation-exchanger within the interval of $CaCl_2$ neutral solution of 42 -176 $mg{\cdot}L^{-1}$ under different temperatures [35]. The Langmuir model was found to be the most suitable to analyze the isotherms. The highest capacity of monolayer has been found to reach 43 mg per 1 g of dry resin at 40º C, it means a half of total exchange capacity is realized. The ΔG, ΔH and ΔS thermodynamic parameters were determined; they show spontaneous endothermic exchange. For the direct $Ca^{2+}{\rightarrow}H^+$ ion exchange, the selectivity coefficient is 3.9, this value is lower for Mg^{2+} ions (2.5) [36]. Investigation of Ca^{2+} and Mg^{2+} sorption on Purolite C100E resin under batch conditions was performed in [37]. Modeling chloride solutions and natural groundwater were applied to the study. The resin dosage, at which no change of the resin capacity occurs, is 4 g l^{-1}. The removal degree of hardness ions from groundwater reaches 70%.

The rate of Ca^{2+} sorption with a Dowex HCR-S /S resin under dynamic conditions was studied in [38]. As shown, increasing (i) the superficial velocity of the solution, (ii) the concentration of the feeding solution from 100 to 1000 ppm, and (III) the grain size causes a decrease of sorption rate. Alternatively, increasing the velocity of flow through K-8H resin accelerates sorption [39]. These different data are evidently caused by various experimental conditions. As a result, the rate determining stages of sorption are different.

3.2   Weakly acidic resins

A number of weakly acidic ion exchangers, for instance, Dowex MAC-3 (Dow Chemical), are based on polyacrylic-divinylbenzene matrix, which contains carboxylic groups (see Fig. 3). The buffer effect caused by them is seen in Fig. 4. The resins are suitable for neutral and alkaline solutions. Thus, they are applied to softening of water, the pH of which is 7 and higher, their usage is ineffective for weakly acidic water.

A Pure PC200FD weakly acidic resin (Pure Resin Co.) was investigated in [37] and compared with strongly acidic resin Purolite C100E. No sufficient difference in exchange capacity towards Ca^{2+} ions has been found, when sorption was performed from a one-component solution. Slightly higher removal degree was realized, when both Ca^{2+} and Mg^{2+} ions were removed from groundwater. FTIR spectra showed no shifts of stripes, which are attributed for carboxylic functional groups (1600 – 1700 cm^{-1}). This indicates no complex formation both for strongly and weakly acidic resins. No sufficient effect of types of anions (Cl$^-$, NO$_3^-$, SO$_4^{2-}$) on calcium uptake by weakly acidic resin is suggested [40].

New commercially available weakly acidic resins were tested in [41] under batch and dynamic conditions. Unfortunately no commercial designation of the resins is specified. As found, the static capacity towards Ca^{2+} ions reaches 70-120 mg g^{-1} under batch conditions. The column test showed high run time (1200-1700 bed volume) in the case of desalination of water containing (mg l^{-1}): Ca^{2+} (20), Mg^{2+} (2), Na$^+$ (4300). In other words, soft water containing a large amount of sodium ions was treated. The hardness ions were removed practically completely: their residual concentration was <1 mg l^{-1} in the effluent. At the same time, a decrease of Na$^+$ was 0.16-0.28 % indicating high selectivity of tested resins towards hardness ions. Higher selectivity towards Ca^{2+} ions compared with Mg^{2+} is stressed.

Ca^{2+} sorption on the Na-form of Dowex MAC-3 resin under batch conditions was studied in [42]. The sodium-loaded resin was obtained by a treatment of the H-form of the resin with a NaOH solution. The isotherms, which were obtained in solutions of a wide concentration interval, are related to the "L" or "H" types of isotherms depending on the concentration of Ca^{2+} ions in equilibrium solution. If the content of these ions is lower than ≈80 mg l^{-1}, the isotherm shows a plateau, further a growth of ion exchange capacity is observed. This phenomenon is explained as the effect of residual content of NaOH, which remained in the resin grains after the transformation of H-form into Na-form. The residual alkali provides an increase of the solution pH inside grains enhancing exchange capacity of weakly acidic resin. The isotherms are fitted by the Langmuir–Vageler model.

Weakly acidic cation exchange resin Lewatit S 8528 (Bayer) has been proposed as an alternative to strongly acidic resins for Ca^{2+} removal even from sugar beet juice [43].

3.3   Polymer-inorganic resins

The usage of polymer-inorganic resins [44-54] are directed to the removal of toxic ions (Ni^{2+} [44,47,49,50], Cd^{2+} [47,48], UO$_2^{2+}$ [46], cationic dye [50]) from water containing also Mg^{2+} and Ca^{2+} ions. However, the hardness ions are removed simultaneously. Thus, the composites are suitable for water softening. Anion exchange resins remove arsenite [52] and chromate [53] anions from the solution containing also an excess of Cl$^-$ or SO$_4^{2-}$ anions.

Strongly acidic Dowex HCR-S [44-50] and weakly acidic Dowex MAC-3 [50] cation exchange resins were applied after modifying with inorganic constituents. Zirconium hydrophosphate was used as a modifier, since its functional groups were able to form complexes of multicharged ions being sorbed [54].

A very important problem is the modifier morphology. It is determined by the precipitation conditions (from deposition of zirconium hydroxophosphate from the $ZrOCl_2$ solution or from sol of hydrated zirconium hydroxocomplexes) and porous structure of polymer matrix. As described in reviews [55,56], porous structure of ion exchange polymers occurs as a result of swelling. It is formed by hydrophilic and hydrophobic voids. Micropores and partially mesopores are hydrophilic, they are caused by the fragments of polymer chains, which contain functional groups. The fragments that are free from them are responsible for the formation of hydrophobic meso- and macropores. Structure defects are also attributed for ion exchange polymers, their size is several tens micrometers.

Thus, the modifier particles of different morphological features can be formed. They can be precipitated in the form of non-aggregated nanoparticles, their aggregates and agglomerates (Fig. 5). Depending on the size, they can be located in different types of pores. Thermodynamic approach to deposition of the particles of one or other size is reported in [57].

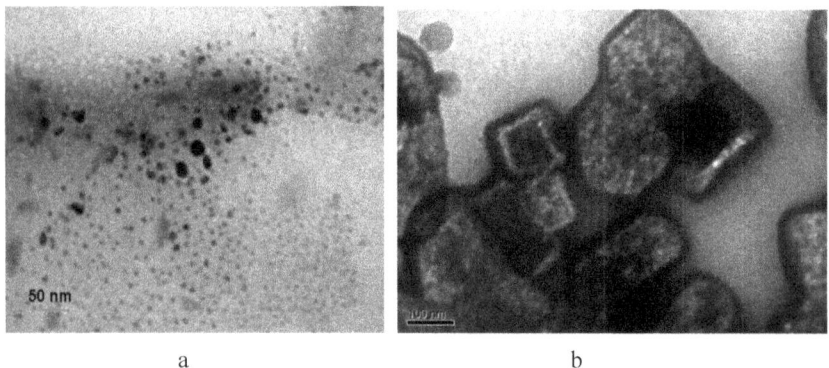

a	b

Figure 5. Non-aggregated (a [45]) and aggregated (b [47])nanoparticles of zirconium hydrophosphate in strongly acidic resin. Reproduced according to the permission of Springer (a) and Elsevier publishers.

The location of inorganic constituents affects properties of composites. Nanoparticles enhance sorption of Ca^{2+}ions [48], as sorption isotherms show (Fig. 6). They are fitted with the Langmuir model. In other words, the curves involve a rapid growth followed by a plateau. At the same time, no sufficient improvement of Ca^{2+} sorption is observed, when the modifier is in the form of large aggregates that are located in hydrophobic pores [47]. These phenomena are explained from the point of view of osmotic pressure of counter- and fixed ions of the modifier. When it is located in hydrophobic pores, the osmotic pressure squeezes hydrophilic pores of the polymer matrix. This is suggested based on porosimetry measurements. As a result, some hydrophilic pores are excluded from ion exchange. The selectivity towards Ni^{2+} and Cd^{2+} ions is due to the inorganic constituents. When the modifier nanoparticles are placed in hydrophilic pores [48], they evidently form additional centers of selective sorption . These centers can involve phosphate and sulfo groups simultaneously. Other advantages of the composites containing nanoparticles are the reproducibility of their composition [51]. Thus, the composite ion exchangers containing non-aggregated nanoparticles of the modifier are the most suitable for water softening.

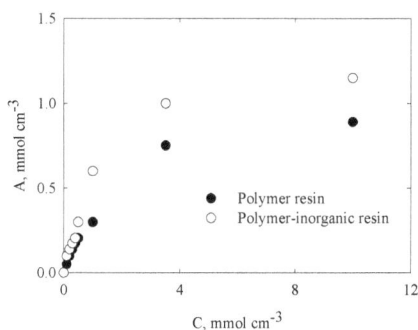

Figure 6. Isotherms of Ca^{2+} sorption on polymer and polymer-inorganic strongly acidic resin. Plotted based on the data of [48].

Polymer-inorganic strongly basic anion-exchanger containing magnetic particles (MIEX resin, Orica Watercare) was mixed with an Amberlite 200C strongly acidic cation exchanger (Dow Chemical) [58]. The cation exchanger contained no inorganic modifier. Unfortunately the modifier morphology as well as the composition of the ion-exchanger were not reported. The tested hard and very hard water contained 300–1000 mg l^{-1} of hardness ions (relatively $CaCO_3$) and 2.9–56 mg l^{-1} of dissolved organic carbon. The ion exchange technique allowed one to reach 97 % removal of total hardness as well as 76 % of organic carbon. As suggested, Ag nanoparticles were embedded to strongly acidic cation exchangers to provide their antimicrobial activity during water softening [59]. However, the suggestion about morphological features of the silver particles was not approved.

4.   Regeneration of ion exchange resins and their fouling

According to the recommendation of the producing company, following reagents are needed for regeneration of resins: H_2SO_4 (1-8 %), HCl (4-8%) or NaCl (8-12%). Water softening requires a solution of NaCl to avoid the formation of acidic effluent. Moreover, this reagent is suitable to regenerate a mixed bed of cation exchange resin and anion exchange resin [58]. The problem is fouling of the resins with organic compounds during water softening. The loading of ion-exchangers with organic substances decreases the efficiency of the removal of hardness ions. As mentioned above, anion exchange resins contain magnetic particles. It means a simple procedure is required for the separation of cation- and anion exchangers. However, it is possible to regenerate the mixed bed without the resin separation, A 2% NaCl solution has been recommended to remove organic compounds from the resins over regeneration. The composition of the waste solution involves high amounts of NaCl, calcium, magnesium, organic compounds and trace contaminants. $CaSO_4$ is precipitated in the waste concentrate, this solid could be removed by filtration. Coagulation or membrane processes are recommended for the recovery of dissolved salts. After the treatment, the regenerating solution can be used repeatedly. The method of the treatment of brine, which is formed after water softening, has been proposed in [60]. The technique involves nanofiltration followed by double-stage crystallization and distillation. A concentrated solution is formed over distillation, it can be used for the regeneration of ion exchange resins.

In comparison with strong acids, the disadvantage of regenerating neutral NaCl solution is that it is impossible to prevent uncontrolled reproduction of microorganisms in the ion exchange column. As found, the microbial composition of water becomes worse after passage through the bed of strongly acidic resin: the amount of microorganisms increases several times [61]. However, it should be noted that the regeneration with back-wash streams depresses them in softened water. This is due to the mechanical destruction of biofilm on the resin grains.

At the same time, it is suggested that the resin loaded with microorganisms cannot be sanitized by flushing, backflushing or other rinsing procedures [62]. Regeneration removes the microorganisms only partially, but it is not enough for sanitation. Recirculation of the regenerating solution promotes fermentation providing bacterial growth. Addition of silver-containing solution to the regenerating liquid depresses bacterial growth. It is necessary to avoid loading of ion-exchangers with silver cations, i.e. to use only low-concentrated solutions. Another way is the disinfection with 0.01% peracetic acid, the combination of silver and peracetic techniques gives better results. As mentioned above, embedding the particles of metal Ag to the resins provides their antimicrobial activity [59].

The regeneration of ion-exchangers is also complicated by the formation of a film, which consists of humic matters, HM [63]. The main reactions, which occur during water softening, are assumed: (i) Na^+ ions in resins are replaced with Ca^{2+} ions, (ii) Ca^{2+} ions in a solution interact with HM, positively charged $HM–Ca^+$ species are formed, (iii) these species are adsorbed by the resin, weakening the electrostatic interaction between Ca^{2+} ions and functional groups of the ion-exchanger. According to the experience of the author of this chapter, this film could be removed only with a solution of strong acid. On the other hand, inorganic modifiers embedded to resin grains can prevent fouling of ion exchange resins [51]. This is an especially valuable advantage of polymer-inorganic ion-exchangers, when they are applied to desalination of biological liquids.

5.   Ion exchange in a combination with other processes

5.1   Ion exchange and ultrasound

Ultrasound is a longitudinal wave, a frequency of which is above 20 kHz [64]. This value is higher than the sonic range (20 Hz - 20 kHz). Humans can hear sounds namely of this frequency diapason. Ultrasound causes cavitation: this process involves formation, increase and collapse of cavities or microbubbles. These processes occur during milliseconds. During the bubble collapse, energy is released in accordance with the theory of "hot spots". The pressure reaches up to 500-10,000 bar, the temperature can reach up to 3000- 5000 K. Under these tough and extreme conditions, the radicals of hydroxyl and hydrogen are generated as a result of thermal dissociation of oxygen and water. These radicals occur in water, oxidizing organic compounds preventing fouling of ion-exchange resins, which is realized, for instance, according to the mechanism [63]. In order to prevent the fouling of polymer materials, ultrasound can be applied to water disinfection, removal of algae, and membrane filtration [64]. It can also be used to decrease the turbidity and total suspended solids.

The application of ultrasound for water softening is considered in [65]. Strongly acidic cation exchanger was used for sorption of Ca^{2+} and Mg^{2+} ions under batch conditions. As confirmed by FTIR spectroscopy, no effect of ultrasound on functional groups occurs. The amount of the ion-exchanger is sufficient: the ultrasonic treatment enhances sorption, when the resin amount is small. Increase in the dosage deteriorates sorption affected by ultrasound, this is probably by the screening of grain surface with rather large bubbles, which cannot be removed from the resin bed. Increase of temperature from 35° to 55° C slightly improves sorption: the growth of the capacity of the Langmuir monolayer is from 26.3 to 27.7 mg g^{-1} towards Mg^{2+} ions (ultrasound), and from 23.2 to 27 mg g^{-1} (no ultrasound). The limiting stage of sorption is the reaction of pseudo-first order. As found, ultrasonic activation accelerates sorption.

Unfortunately, the disadvantage of this technique is a strong dependence of the hardness removal efficiency on the resin dosage. This makes impossible the usage of ultrasound for ion exchange columns.

5.2   Ion exchange and electrodialysis

The electrochemical method, which combines ion exchange and electrodialysis, is called "electrodeionization", EDI. It consists of two main stages: (i) extraction of ions by ion-exchange resin, (ii) migration of species being sorbed through the ion exchanger and membrane towards the electrode compartment of the electrodialysis cell (Fig. 7). The bed of ion-exchanger (cation, or anion exchange resins, or their mixture) is placed between cation and anion exchange membranes. As opposed to traditional ion exchange, EDI can be considered a continuous process. The necessary condition of continuity is the equality of rates of sorption and migration [66]. Regarding divalent ions, a high rate of their migration can be achieved only in flexible resins. Different fields of the EDI application are considered in reviews [67-69].

It should be noted that the EDI method is the most suitable for water containing no hardness ions. When ions, which are present in water, form insoluble compounds, the EDI process is complicated with deposit formation on the surface of both resin grains and membranes, since water dissociation occurs on the solid-liquid interface. H^+ ions enter the ion-exchanger bed, OH^- ions remain at the surface of cation exchange grains and separator. Precipitation is a result of the solution alkalization. In order to solve this problem, acidification of feeding solution has been proposed [70]. Acidic solution was also used in concentrating compartments of the EDI cell. Flexible Dowex WX-2 resin was applied to the EDI process. However, leakage of acid through the membranes (especially through anion exchange separator) is possible, as a result, the solution being desalinated is acidified. However, this approach is not suitable for the softening of water, since its neutrality must be provided.

In order to prevent the deposit formation, periodic change of the polarity of the EDI cell (electrodeionization reversal, EDIR) has been proposed [70]. A mixed bed of the Amberlite IRA 402 Cl and Amberlite IR 120 Na resins, which were produced by Rohm and Hass Co. (USA)б was placed between membranes. The model solutions, the hardness of which varied from 250 to 1250 mg l^{-1} relative to $CaCO_3$, were passed through the resin bed. The removal degree of calcium and magnesium ions exceeded 99 %, when the period of polarity reversal was 10 min. This approach allowed one to perform a long-time operation of the EDIR system and to minimize scale formation (the deposit contains mainly $CaCO_3$ and

MgCO$_3$). However, water was acidified down to pH 4 during the passage from the bed indicating preferable removal of cations. The removal degree of hardness ions reached about 20% during the EDIR process [72]. However, the energy consumption was ≈20 kWh per 1 kg of CaCO$_3$ indicating low economical efficiency of the process. For instance, in the case of electrodialysis, the energy consumption is 1 kWh per 1 kg of salts.

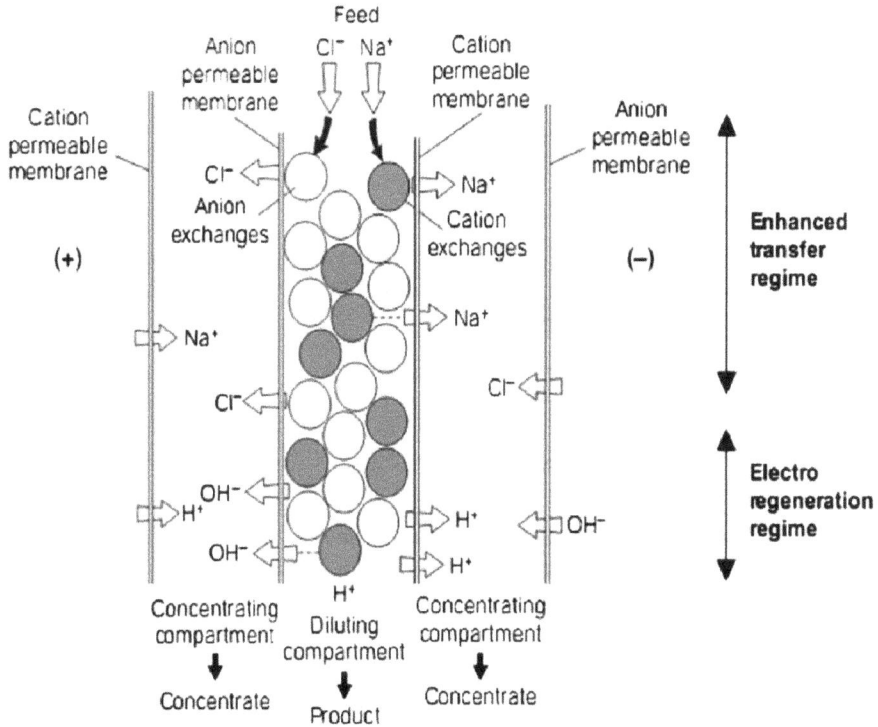

Figure 7. Scheme of the EDI process [67]. Reproduced according to the permission of Elsevier Publisher.

Conclusions

Among various methods of water softening, ion exchange possesses a special position, since it requires no expensive and complex materials and equipment. A wide range of strongly and weakly acidic cation-exchange resins are proposed by producing companies.

Thus, the ion-exchange resins are easily accessible. The mentioned advantages of ion exchange techniques allows one to use them for various application fields despite discontinuity of these processes. As opposed to traditional ion exchange, continuous water softening (EDI, EDIR) requires high energy consumption.

Ion exchange resins are easily regenerated with NaCl solution, the brine is eco-friendly. Fouling and biofouling, which occur over water softening and resin regeneration, can be prevented by modifying resins with inorganic ion-exchangers or silver particles. Economical aspects of the usage of composite resins should be considered for each case.

References

[1] E. Yildiz, A. Nuhoglu, B. Keskinler, G. Akay, B. Farizoglu, Water softening in a crossflow membrane reactor, Desalination 153 (2003) 139-152. https://doi.org/10.1016/S0011-9164(03)90066-X

[2] J. Liorens, J. Sabaté, M. Pujola, Viability of the use of polymer-assisted Ultrafiltration for continuous water softening, Sep. Sci. Technol. 38 (2003) 295-322. https://doi.org/10.1081/SS-120016576

[3] Yu. Dzyazko, L. Rozhdestveska, V. Ogenko, Yu. Borysenko, A. Bildukevich, T. Plisko, Y. Zmievskii, Polymer-inorganic membranes modified with graphene-containing composite: Electrochemical approach to investigations of functional properties, Mater. Today Proc. 50 (2022) 507-523. https://doi.org/10.1016/j.matpr.2021.11.303

[4] P. Jin, M. Robeyn, J. Zheng, S. Yuan, B. van der Bruggen, Tailoring charged nanofiltration membrane based on non-aromatic Tris (3-aminopropyl) amine for effective water softening, Membranes. 10 (2020) 251. https://doi.org/10.3390/membranes10100251

[5] X. Li, D. Hasson, R. Semiat, H. Shemer, Intermediate concentrate demineralization techniques for enhanced brackish water reverse osmosis water recovery - A review, Desalination 466 (2019) 24-35. https://doi.org/10.1016/j.desal.2019.05.004

[6] R. Zadghaffari, S.S. Asr, Water softening using caustic soda: privileges and restrictions, Polish J. Chem. Technol. 15 (2013) 116-121. https://doi.org/10.2478/pjct-2013-0033

[7] M. Ostovar, M. Amiri, A novel eco-friendly technique for efficient control of water lime softening process, Water Env. Res. 85 (2013) 2285-2293. https://doi.org/10.2175/106143013X13807328848333

[8] D. Campanizzi, B. Mason, C.K.F. Hermann, Distillation apparatuses Using household items, J. Chem. Educ. 76 (1999) 1079-1080. https://doi.org/10.1021/ed076p1079

[9] A. Abdel-Karim, S. Leaper, C. Skuse, G. Zaragoza, Membrane cleaning and pretreatments in membrane distillation - a review, Chem. Eng. J. 422 (2021) 129696. https://doi.org/10.1016/j.cej.2021.129696

[10] P.K. Naryanan, W.P. Harkare, S.K. Adhikary, N.J. Dave, D.K. Chauhan, K.P. Govindan, Performance of an electrodialysis desalination plant in rural areas, Desalination. 54 (1985) 145-150. https://doi.org/10.1016/0011-9164(85)80013-8

[11] J. Choi, P. Dorji, H.K. Shon, S. Hong, Applications of capacitive deionization: Desalination, softening, selective removal, and energy efficiency, Desalination. 445 (2019) 118-130. https://doi.org/10.1016/j.desal.2018.10.013

[12] L. Wang, S. Lin, Mechanism of selective ion removal in membrane capacitive deionization for water softening, Environ. Sci. Technol. 53 (2019) 5797-5804. https://doi.org/10.1021/acs.est.9b00655

[13] I. Sanjuán, D. Benavente, E. Expósito, V. Montiel, Electrochemical water softening: Influence of water composition on the precipitation behavior, Sep. Purif. Technol. 211 (2019) 857-865. https://doi.org/10.1016/j.seppur.2018.10.044

[14] K. Yaghmaeian, A. Mahvi, S. Nasseri, M. H. Shayesteh, H. J. Mansoorian, N. Khanjani, Drinking water softening with electrocoagulation process: Influence of direct and alternating currents as inductive with different arrangement rod electrodes and polarity inverter, Scientia Iranica. 27 (2021) 1275-1292.

[15] M. Micari, M. Moser, A. Cipollina, A. Tamburini, G. Micale, V. Bertsch, Towards the implementation of circular economy in the water softening industry: A technical, economic and environmental analysis, J. Cleaner Product. 255 (2020) 120291. https://doi.org/10.1016/j.jclepro.2020.120291

[16] World Health Organization, Hardness in Drinking-Water. Background document for development of WHO Guidelines for Drinking-Water Quality, 2011.

[17] S. Baskar, R. Baskar, L. Mauclaire, J. A. McKenzie, Microbially induced calcite precipitation in culture experiments: Possible origin for stalactites in Sahastradhara caves, Dehradun, India, Current Sci. 90 (2006) 58-64.

[18] L.F. Capitan-Vallvey, M.D. Fernandez-Ramos, P.A. De Cienfuegos Galvez, F. Santoyo-Gonzalez, Characterisation of a transparent optical test strip for quantification of water hardness, Anal. Chim. Acta. 481 (2003) 139-148. https://doi.org/10.1016/S0003-2670(03)00073-4

[19] J. Saurina, E. López-Aviles, A. Le Moal, S. Hernández-Cassou, Determination of calcium and total hardness in natural waters using a potentiometric sensor array, Anal. Chim. Acta. 464 (2002) 89-98. https://doi.org/10.1016/S0003-2670(02)00474-9

[20] M.I.S.Veríssimo, J.A.B.P. Oliveira, M.T.S.R. Gomes, Determination of the total hardness in tap water using acoustic wave sensors, Sens. Actuators B Chem. 127 (2007) 102-106. https://doi.org/10.1016/j.snb.2007.07.006

[21] L.F.Capitán-Vallvey, M.D. Fernández-Ramos, P.A. De Cienfuegos Gálvez, F. Santoyo-González, Characterisation of a transparent optical test strip for quantification of water hardness, Anal. Chim. Acta. 481(2003) 139-148. https://doi.org/10.1016/S0003-2670(03)00073-4

[22] M.L. Bouhoun, P. Blondeau, Y. Louafi, F.J. Andrade, A Paper-Based Potentiometric Platform for Determination of Water Hardness, Chemosensors. 9 (2021) 96. https://doi.org/10.3390/chemosensors9050096

[23] M. Shariati-Rad, S. Heidari, Classification of and determination of total hardness of water using silver nanoparticles, Talanta. 219 (2020) 121297. https://doi.org/10.1016/j.talanta.2020.121297

[24] Gray N.F. Drinking water quality problems and solutions, second ed., Cambridge,University Press, 1994.

[25] P. Sengupta, Potential Health Impacts of Hard Water. Int. J. Prev. Med. 4 (2013) 866-875.

[26] E. Rubenowitz-Lundin, K.M. Hiscock, Water Hardness and Health Effects. In: O.e.a. Selinus (Eds.), Essentials of Medical Geology, Revised Edition, Springer, Dordrecht, Heidelberg, New York, London, 2013. https://doi.org/10.1007/978-94-007-4375-5_14

[27] K. Valta, T. Kosanovic, D. Malamis, K. Moustakas, M. Loizidou, Overview of water usage and wastewater management in the food and beverage industry, Desalination and Water treatment, 53 (2015) 3335-3347. https://doi.org/10.1080/19443994.2014.934100

[28] R.A. Ward, Worldwide water standards for hemodialysis, Hemodialysis International, 11 (2007) S18-S25. https://doi.org/10.1111/j.1542-4758.2007.00142.x

[29] K. Bensadok, A.Refes, P.M.Charvier, G.Nezzal, Water produce for pharmaceutical industry: role of reverse osmosis state, Desalination. 22 (2008) 298-302. https://doi.org/10.1016/j.desal.2007.01.086

[30] A. Auron, U.S. Alon, Hypercalcemia: a consultant's approach, Pediatric Nephrology. 33 (2018) 1475-1488. https://doi.org/10.1007/s00467-017-3788-z

[31] V. Bellizzi, L. De Nicola , Minutolo R, Russo D, Cianciaruso B, Andreucci M, et al. Effects of water hardness on urinary risk factors for kidney stones in patients with idiopathic nephrolithiasis, Nephron. 81 (1999) 66-70. https://doi.org/10.1159/000046301

[32] A.K. Chandra, P. Sengupta, H. Goswami, M. Sarkar, Excessive dietary calcium in the disruption of structural and functional status of adult male reproductive system in rats with possible mechanism, Mol. Cell Biochem. 364 (2012) 181-191. https://doi.org/10.1007/s11010-011-1217-3

[33] P. Sengupta, The laboratory rat: Relating its age with human's, Int. J. Prev. Med. 4 (2013) 624-630.

[34] H. Hoffmann, F. Martinola, Selective resins and special processes for softening water and solutions; A review, Reactive Polymers, Ion Exchangers, Sorbents. 7 (1988) 263-272. https://doi.org/10.1016/0167-6989(88)90148-1

[35] B. Bandrabur, R.-E. Tataru-Fărmuş, L. Lazăr, G. Gutt, Application of a strong acid resin as ion exchange material for water softening - equilibrium and thermodynamic analysis, Scholarly J. 13 (2012) 361-370.

[36] W. S. Miller, C. J. Castagna, A. W. Pieper, Understanding Ion-Exchange Resins For Water Treatment Systems, GE Water and Process Technologies. (2009) 1-13.

[37] B. Bandrabur, L. Lazar, R.-E. Tataru-Farmus, L. Bulgariu, G. Gutt, Permanent hard water softening using different cation exchange resins, Buletinul Institutului Politehnic in Jasi LVIII (2012) 141-150.

[38] N. N. Ismail, Experimental study on ion exchange rate of calcium hardness in water softening process using strong acid resin DOWEX HCR S/S, J. Eng. Sci. 19 (2016) 107-114.

[39] A.A.Swelam, A.M.A. Salem, M.B Awad, Permanent Hard Water Softening Using Cation Exchange Resin in Single and Binary Ion Systems, World J. Chem. 8 (2013) 1-10.

[40] B. Bandrabur, L. Lazar, R.-E. Tataru-Farmus, G. Gutt, Cationic exchange capacity of PURE PC200FD resin in food industry water softening process, Buletinul Institutului Politehnic in Jasi XI (2012) 97-102.

[41] A. Janson, J. Minier-Matar, E. Al-Shamari, A. Hussain, R. Sharma, D. Rowley, Evaluation of new ion exchange resins for hardness removal from boiler feedwater, Emergent Mater. 1 (2018) 77-87. https://doi.org/10.1007/s42247-018-0006-0

[42] G.J. Millar, S. Papworth, S.J. Couperthwaite, Exploration of the fundamental equilibrium behaviour of calcium exchange with weak acid cation resins, Desalination. 351 (2014) 27-36. https://doi.org/10.1016/j.desal.2014.07.022

[43] M. Coca, S. Mato, G. Gonzalez-Benito, M. A. Uruena, M. T. Garcia-Cubero, Use of weak cation exchange resin Lewatit S 8528 as alternative to strong ion exchange resins for calcium salt removal, J. Food Eng. 97 (2010) 569-573. https://doi.org/10.1016/j.jfoodeng.2009.12.002

[44] Y.S. Dzyazko, L.N. Ponomareva, Y.M. Volfkovich, V.E. Sosenkin, Effect of the porous structure of polymer on the kinetics of Ni2+ exchange on hybrid inorganic-organic ionites, Russ. J. Phys. Chem. A 86((2012) 913-919. https://doi.org/10.1134/S0036024412060088

[45] Y.S. Dzyazko, L.N. Ponomareva, Y.M. Volfkovich, V.E. Sosenkin, V.N. Belyakov, Conducting properties of a gel ionite modified with zirconium hydrophosphate nanoparticles. Russ. J. Electrochem. 49 (2013) 209-215. https://doi.org/10.1134/S1023193513030075

[46] Y.S. Dzyazko, O.V. Perlova, N.A. Perlova, Y.M. Volfkovich, V.E. Sosenkin, V.V. Trachevskii, V.F. Sazonova, A.V. Palchik, Composite cation-exchange resins containing zirconium hydrophosphate for purification of water from U(VI) cations, Desalination Water Treatment. 69 (2017) 142-152. https://doi.org/10.5004/dwt.2017.0686

[47] Y.S. Dzyazko, L.N. Ponomaryova, Y.M. Volfkovich, VV Trachevskii, A.V. Palchik, Ion-exchange resin modified with aggregated nanoparticles of zirconium hydrophosphate. morphology and functional properties, Micropor. Mesopor. Mater. 198 (2014) 55-62. https://doi.org/10.1016/j.micromeso.2014.07.010

[48] Y.S. Dzyazko, L.N. Ponomaryova, Y.M. Volfkovich, V.E. Sosenkin, Polymer ion-exchangers modified with zirconium hydrophosphate for removal of Cd2+ ions from diluted solutions, Separ. Sci. Technol. 48 (2013) 2140-2149. https://doi.org/10.1080/01496395.2013.794434

[49] Yu. Dzyazko, L. Ponomarova, Yu. Volfkovich, V. Tsirina, V. Sosenkin, N. Nikolska, V. Belyakov, Influence of zirconium hydrophosphate nanoparticles on porous structure and sorption capacity of the composites based on ion exchange resin,

Chemistry and Chemical Technology. 10 (2016) 329-335. https://doi.org/10.23939/chcht10.03.329

[50] L. Ponomareva, Y. Dzyazko, Y. Volfkovich, V. Sosenkin, S. Scherbakov, Effect of Incorporated Inorganic Nanoparticles on Porous Structure and Functional Properties of Strongly and Weakly Acidic Ion Exchangers, Springer Proc. Phys. 214 (2017) 63-77. https://doi.org/10.1007/978-3-319-92567-7_4

[51] Yu. Dzyazko, Yu. Borysenko, Yu. Zmievskii, V. Zakharov, V. Myronchuk, E. Kolomiets, Organic-inorganic ion exchange materials for electromembrane processing of liquid wastes produced in the dairy industry, Mater. Today: Proc. 50 (2022) 496-501. https://doi.org/10.1016/j.matpr.2021.11.301

[52] T.V. Maltseva, E.O. Kolomiets, Y.S. Dzyazko, S. Scherbakov, Composite anion-exchangers modified with nanoparticles of hydrated oxides of multivalent metals, Appl. Nanosci. 9 (2019) 997-1004. https://doi.org/10.1007/s13204-018-0689-9

[53] Y. Dzyazko, E. Kolomyets, Y. Borysenko, V. Chmilenko, I. Fedina, Organic-inorganic sorbents containing hydrated zirconium dioxide for removal of chromate anions from diluted solutions, Mater. Today: Proc. 6 (2019) 260-269. https://doi.org/10.1016/j.matpr.2018.10.103

[54] Y.S. Dzyazko, V.V. Trachevskii , L.M. Rozhdestvenskaya, S.L. Vasilyuk V.N. Belyakov, Interaction of sorbed Ni(II) ions with amorphous zirconium hydrogen phosphate. Russ. J. Phys. Chem. A 87((2013) :840-845. https://doi.org/10.1134/S0036024413050063

[55] Y. Dzyazko, Y. Volfkovich, O. Perlova, L. Ponomaryova, N. Perlova, E. Kolomiets, Effect of Porosity on Ion Transport Through Polymers and Polymer-Based Composites Containing Inorganic Nanoparticles (Review), Springer Proc. Phys. 222 (2019) 235-253. https://doi.org/10.1007/978-3-030-17755-3_16

[56] Y. Dzyazko, A. Omel'chuk, Porous ionic polymers, in: Inamuddin, M. I. Ahamed, R. Boddula (Eds.) Porous Polymer Science and Application, CRC Press, Boca Raton, 2022, pp. 37-59. https://doi.org/10.1201/9781003169604-3

[57] N. Perlova, Y. Dzyazko, O. Perlova, A. Palchik, V Sazonova, Formation of zirconium hydrophosphate nanoparticles and their effect on sorption of uranyl cations, Nanoscale Res. Let. 12 (2017) 209. https://doi.org/10.1186/s11671-017-1987-y

[58] S.E.H. Comstock, T.H. Boyer, Combined magnetic ion exchange and cation exchange for removal,Chem. Eng. J. 241 (2014) 366-375. https://doi.org/10.1016/j.cej.2013.10.073

[59] R. Khaydarov, M. Abdukhakimov, I. Garipov, I. Sadikov, P. T. Krishnamurthy, S. Evgrafova, Silver-containing cation exchange resin: synthesis and application, Mater. Sci. (Medziangotyra). 28 (2022) 89-92. https://doi.org/10.5755/j02.ms.28473

[60] M. Micari, A. Cipollina, A. Tamburini, M. Moser, V. Bertsch, G. Micale, Combined membrane and thermal desalination processes for the treatment of ion exchange resins spent brine, Appl. Energy 254 (2019) 113699. https://doi.org/10.1016/j.apenergy.2019.113699

[61] S.A. Parsons, The effect of domestic ion-exchange water softeners on the microbiological quality of drinking water, Wat. Res. 34 (2000) 2369-2375. https://doi.org/10.1016/S0043-1354(99)00407-8

[62] H.-C. Flemming, Microbial growth on ion exchangers, Wat. Res. 21 (1987) 745-756. https://doi.org/10.1016/0043-1354(87)90149-7

[63] B. Dong, Y. Xu, D. Shen, X. Dai, S. Lin, Characterizing the interactions between humic matter and calcium ions during water softening by cation-exchange resins, RSC Adv. 6 (2016) 93947. https://doi.org/10.1039/C6RA22113K

[64] M.R. Doosti, R. Kargar, M.H. Sayadi, Water treatment using ultrasonic assistance: A review, Proceedings of the International Academy of Ecology and Environmental Sciences, 2012, 2 (2) 96-110.

[65] M.H. Entezari, M. Tahmasbi, Water softening by combination of ultrasound and ion exchange, Ultrasonics Sonochemistry. 16 (2009) 356-360. https://doi.org/10.1016/j.ultsonch.2008.09.008

[66] Yu.S. Dzyatsko, L.N. Ponomareva, L.M. Rozhdestvenskaya, S.L. Vasilyuk. V.N. Belyakov, Electrodeionization of low-concentrated multicomponent Ni2 +-containing solutions using organic-inorganic ion-exchanger, Desalination. 342 (2014) 43-51. https://doi.org/10.1016/j.desal.2013.11.030

[67] Ö. Arar, Ü. Yüksel, N. Kabay, M. Yüksel, Various applications of electrodeionization (EDI) method for water treatment-A short review, Desalination. 342 (2014) 16-22. https://doi.org/10.1016/j.desal.2014.01.028

[68] B.S. Rathi, P.S. Kumar, R. Parthiban, A review on recent advances in electrodeionization for various environmental applications, Chemosphere. 289 (2022) 133223. https://doi.org/10.1016/j.chemosphere.2021.133223

[69] B.S. Rathi, P.S. Kumar, Electrodeionization theory, mechanism and environmental applications. A review, Environ. Chem. Letter. 18 (2020) 1209-1227. https://doi.org/10.1007/s10311-020-01006-9

[70] P.B. Spoor, L. Grabovska, L. Koene, L.J.J. Janssen, W.R. ter Veen, Pilot scale deionisation of a galvanic nickel solution using a hybrid ion-exchange/electrodialysis system, Chem. Eng. J. 89 (2002) 193-202. https://doi.org/10.1016/S1385-8947(02)00009-8

[71] H.J. Lee, M.K. Hong, S.H. Moon, A feasibility study on water softening by electrodeionization with the periodic polarity change, Desalination. 284 (2012) 221-227. https://doi.org/10.1016/j.desal.2011.09.001

[72] H. Jin, Y. Yu, L. Zhang, R. Yan, X. Chen, Polarity reversal electrochemical process for water softening, Sep. Pur. Technol. 210 (2019) 943-949. https://doi.org/10.1016/j.seppur.2018.09.009

Keyword Index

Affinity Chromatography	39
Alkaloid	75
Anionic Resins	120
Bioactive Molecules	75
Biochemical Purification	75
Buffer Solution	39
Carbohydrates	75
Catechin	75
Cation and Anion Exchange Chromatography	39
Cationic and Anionic Exchangers	1
Cationic Resins	120
Donnan Equilibrium	39
Drug Release Kinetics	93
Drug Release	120
Effect of Buffer	1
Effect of Support	1
Effective Pollution Control Method	55
Electrodeionization	142
Fat Soluble Vitamins	24
Gel Filtration or Permeation Chromatography	39
Hardness Ions	142
Human C-Peptide	75
Immunoaffinity Chromatography	39
Microencapsulation	93
Peptides	75
Pi of Proteins	1
Plasmid DNA	75
Polymer-Inorganic Resins	142
Polyphenols	75
Polystyrene Divinylbenzene	1
Proteins	75
Purification	24
Resin Characterization	1
Resin Functionalization	1
Resinate	93
Separation	24
Steps for Separation	1
Sustained Drug Delivery	93
Targeted Drug Delivery	120
Therapeutics	120
Toxic Metal	55
Type of Proteins	1
Vitamins	24
Water Softening	142
Water Soluble Vitamins	24

About the Editors

Dr. Inamuddin is working as Assistant Professor at the Department of Applied Chemistry, Aligarh Muslim University, Aligarh, India. He obtained a Master of Science degree in Organic Chemistry from Chaudhary Charan Singh (CCS) University, Meerut, India, in 2002. He received his Master of Philosophy and Doctor of Philosophy degrees in Applied Chemistry from Aligarh Muslim University (AMU), India, in 2004 and 2007, respectively. He has extensive research experience in multidisciplinary fields of Analytical Chemistry, Materials Chemistry, and Electrochemistry and, more specifically, Renewable Energy and Environment. He has worked on different research projects as project fellow and senior research fellow funded by University Grants Commission (UGC), Government of India, and Council of Scientific and Industrial Research (CSIR), Government of India. He has received Fast Track Young Scientist Award from the Department of Science and Technology, India, to work in the area of bending actuators and artificial muscles. He has also received the Sir Syed Young Researcher of the Year Award 2020 from Aligarh Muslim University. He has completed four major research projects sanctioned by University Grant Commission, Department of Science and Technology, Council of Scientific and Industrial Research, and Council of Science and Technology, India. He has published 207 research articles in international journals of repute and nineteen book chapters in knowledge-based book editions published by renowned international publishers. He has published 165 edited books with Springer (U.K.), Elsevier, Nova Science Publishers, Inc. (U.S.A.), CRC Press Taylor & Francis Asia Pacific, Trans Tech Publications Ltd. (Switzerland), IntechOpen Limited (U.K.), Wiley-Scrivener, (U.S.A.) and Materials Research Forum LLC (U.S.A). He is a member of various journals' editorial boards. He is also serving as Associate Editor for journals (Environmental Chemistry Letter, Applied Water Science and Euro-Mediterranean Journal for Environmental Integration, Springer-Nature), Frontiers Section Editor (Current Analytical Chemistry, Bentham Science Publishers), Editorial Board Member (Scientific Reports-Nature) and Review Editor (Frontiers in Chemistry, Frontiers, U.K.) He has also guest-edited various special thematic special issues to the journals of Elsevier, Bentham Science Publishers, and John Wiley & Sons, Inc. He has attended as well as chaired sessions in various international and national conferences. He has worked as a Postdoctoral Fellow, leading a research team at the Creative Research Initiative Center for Bio-Artificial Muscle, Hanyang University, South Korea, in the field of renewable energy, especially biofuel cells. He has also worked as a Postdoctoral Fellow at the Center of Research Excellence in Renewable Energy, King Fahd University of Petroleum and Minerals, Saudi Arabia, in the field of polymer electrolyte membrane fuel cells and computational fluid dynamics of polymer electrolyte membrane fuel cells. He is a life member of the Journal of the Indian Chemical Society. His research interest includes ion exchange materials, a sensor for heavy metal ions, biofuel cells, supercapacitors and bending actuators.

Ms. Maha Khan is a Research Scholar at the Department of Applied Chemistry, Aligarh Muslim University (A.M.U.), Aligarh, India. She has also pursued her Bachelor's in Chemistry and Master's in Polymer Science and Technology from A.M.U., Aligarh. Her research work focuses primarily on Enzymatic Biofuel Cells, a pathway to clean and green energy.

Dr. Mohammad A. Jafar Mazumder has been serving as a Professor of Chemistry at King Fahd University of Petroleum & Minerals (KFUPM), Saudi Arabia. He has extensive experience in designing, synthesizing, and characterizing various organic compounds, ionic and thermo-responsive polymers for corrosion, water treatment, and biomedical applications. Dr. Jafar Mazumder obtained his B.Sc (Hons.), M.Sc (Chemistry) from Aligarh Muslim University, India, MS (Chemistry) from KFUPM, Saudi Arabia, and Ph.D. in Chemistry (2009) from McMaster University, Canada.

In more than 20 years of academic research, Dr. Jafar Mazumder has had the opportunity to work with several international collaborative research groups and has exposed himself to a broad range of research areas. Dr. Jafar Mazumder secured 7 US patents, published more than 85 articles in peer-reviewed journals, 37 conference proceedings, 9 book chapters, and co-edited 4 books with Springers and Trans Tech publications. He is awarded as a Fellow of the Royal Society of Chemistry and Chartered Chemist, Association of Chemical Profession of Ontario, Canada. Besides, Dr. Jafar Mazumder is a member of the American Chemical Society (ACS), Canadian Society for Chemistry (CSC), Canadian Biomaterial Society (CBS), and a life member of Bangladesh Chemical Society (BCS). In his academic career, he was awarded numerous national and international scholarships and awards that include the prestigious Indian Council for Cultural Relations (ICCR) Scholarship from Govt. of India for undergraduate studies in India, Aligarh Muslim University undergraduate & graduate Gold medal, and certificate of excellence from the Ministry of Human Resource Development, Govt. of India, and MITACS postdoctoral fellowship (Canada) for pursuing postdoctoral research in Chemical and Biomedical Engineering.

Currently, Dr. Jafar Mazumder is actively involved in several ongoing university (KFUPM), government (KACST, NSTIP), and client (Saudi Aramco) funded projects in the capacity of principal and co-investigators. His current research interest includes the design, synthesis, and characterization of various modified monomers and polymers for potential use in the inhibition of mild steel corrosion in oil and gas industries and the preparation of multilayered polyelectrolyte coated membranes for the removal of heavy metals and organic contaminants from aqueous water samples. Long term scientific goal of Dr. Jafar Mazumder is not merely to make science fun and entertaining for people. It is to engage them with a multidisciplinary scientific mission at a deeper level to create a space through which they can interact with scientific ideas, develop connections between science, engineering, and biology, and thoughts of their own to contributions to society. He feels this goal and engaging personality make him a pleasant person to work with and help inspire his co-workers in any professional setting.

Dr. Mohammad Luqman has 12+ years of post-PhD experience in Teaching, Research, and Administration. Currently, he is serving as an Assistant Professor of Chemical Engineering in Taibah University, Saudi Arabia. Before joining here, he served as an Assistant Professor in College of Applied Science at A'Sharqiyah University, Oman, and in College of Engineering at King Saud University, Saudi Arabia. He served as a Research Engineer in SAMSUNG Cheil Industries, South Korea. Moreover, he served as a post-doctoral fellow at Artificial Muscle Research Center, Konkuk University, South Korea, in the field of Ionic Polymer Metal Composites for the development of Artificial Muscles, Robotic Actuators and Dynamic Sensors. He earned his PhD degree in the field of Ionomers (Ion-containing Polymers), from Chosun University, South Korea. He successfully served as an Editor to three books, published by world renowned publishers. He published numerous high-quality papers, and book chapters. He is serving as an Editor and editorial/review board members to many International SCI and Non-SCI journals. He has attracted a few important research grants from industry and academia. His research interests include but not limited to Development of Ionomer/Polyelectrolyte/non-ionic Polymer Nanocomposites/Blends for Smart and Industrial/Engineering Applications.

www.ingramcontent.com/pod-product-compliance
Lightning Source LLC
Chambersburg PA
CBHW071234210326
41597CB00016B/2047